MOULT in Birds

H B Ginn & D S Melville

BTO GUIDE 19

BRITISH TRUST FOR ORNITHOLOGY
GUIDE NUMBER NINETEEN

ISBN 0 903793 02 4

Published by
The British Trust for Ornithology,
Beech Grove, Tring, Hertfordshire, England
1983

Printed by Maund & Irvine Ltd., Brook Street, Tring, Herts.

CONTENTS

PREFACE

Moult, in which old feathers are replaced by new, is an important event in the annual cycle of a bird. A knowledge of moult is fundamental to an understanding of the life history, which in turn is necessary before conservation measures can be designed and implemented, for any species.

The Moult Enquiry was started by the British Trust for Ornithology in 1960. One of its original aims was the collection of information from trapped birds on the patterns and sequence of moult, to aid ringers in ageing their catch (Williamson 1960a). It soon became obvious that the study had ramifications far beyond these original objectives, with diverse and fascinating adaptations becoming apparent as to timing, rate, extent, frequency and duration of moult in relation to breeding and migration, features in turn affected by climate and food availability. There is now increasing interest in the effects of weather on populations and breeding success, and moult studies of adults and juveniles can contribute significantly in helping to determine times of fledging and the end of the breeding season. These events are less easily determined through the Nest Records Scheme due to factors such as increased vegetation cover and waning observer interest as the season progresses.

Following the initial success of the Moult Enquiry D. W. Snow prepared "A Guide to Moult in British Birds" (1967, B.T.O. Field Guide No. 11), which reviewed the available information and provided general instructions and advice on collecting moult data. This was reprinted in 1970 without revision but has long been out of print.

The present work has had a lengthy gestation period. Part I, and the moult score figures for Part II were prepared in 1977/78 by H. B. Ginn, assisted by Peter Marsack and David Partridge. The compilation of the species accounts and a partial revision of Part I were carried out by D. S. Melville in 1982. The order of authorship has been determined alphabetically. The work draws heavily on its predecessor and we are most grateful to Dr Snow for permission to do so. The great increase in the information currently available on moult in British birds, as evidenced in Part II, reflects the success of the earlier Guide and we hope that future workers will find the present work of value in their studies. There is still much to be learned about the many aspects of moult in British birds.

AIMS AND ORGANISATION

The aims of this work can be summarised as follows:-

a) to give a brief summary for the ornithologist and interested layman alike, of the nature and variety of plumage and of the fascinating diversity of regimes for periodically renewing it.

b) to provide bird ringers, museum workers and other ornithologists with a basic text on technique for recording moult and details of the sort of moult patterns which they may expect to find.

c) to indicate gaps in our knowledge of the subject with the object of encouraging further moult studies.

The work falls into two parts. The first section gives a short general summary of the plumage of birds, its structure, arrangement and functions and briefly reviews the various types of moult patterns which are found and the ways of recording them. While the text is mainly concerned with European species, examples are given of moult patterns from elsewhere where these help to put the patterns of moult found in Europe (mainly temperate and subarctic) into a wider perspective. The information given is necessarily a very brief summary of a vast literature and in general the only references cited are those to key works giving very full accounts or dealing with particular species (references are not given in Part I for those species included in Part II). Some subjects which are very relevant to the plumage and moult of birds are largely outside the scope of this work. These include the embryology of feathers, the wide variation in fine structure of feathers from different parts of the body and from different species, the function of plumage in display, and detailed accounts of hormonal aspects of moult control. However, some references to detailed accounts of these subjects are also given so that the reader can achieve a complete picture of the topic if desired. For more general accounts of bird biology reference should be made to works by Pettingill (1970), Van Tyne and Berger (1976) and Welty (1975). Voitkevich (1966) gives a detailed account of the plumage of birds.

The second part of the present work gives, in a very condensed form, a systematic species-by-species account of such points as the number of flight feathers, the sequence, season and rate of moult, its relationship to breeding and migration, and the extent of the post-juvenile moult. In addition, scatter diagrams of primary moult score plotted against date are given for some species. These provide a quick visual representation of the annual primary moult Although a wide literature search has been undertaken it is not claimed that the details are in anyway complete, as much information on moult is scattered through the literature, very often as brief mentions in papers dealing with quite different subjects.

A species has been included in Part II if it has bred in Britain in recent years or if it regularly undergoes a major moult (including some or all flight feathers) in Britain (e.g. as in a number of waders). This selection was necessary to reduce the number of species to manageable proportions, though the B.T.O. Moult Enquiry welcomes moult records for all species. In particular, people handling less common birds should examine them for moult as a matter of routine (e.g. a Sooty Tern *Sterna fuscata* found in Northamptonshire in June 1980 and a Rüppell's Warbler *Sylvia ruppelli* caught in Shetland in 1978 were in active primary moult).

Since it is always useful to know where the gaps in our knowledge lie, each species in the systematic list is given a star rating from one to four (beside the scientific name in the heading), giving an approximate indication of the number of moult cards held by the B.T.O. (see page 32). The amount of useful and usable data on cards varies enormously so that the actual number of cards can be only a rough indication of the amount of information available. One and two-star species should be examined for moult on every practicable occasion, whether found dead or caught for ringing, while data are still badly needed for three-star species. Though reasonable samples of cards are available for four-star species, records are still required to permit future comparisons of the timing and duration of moult between different geographical areas and between different years. Information filling any of the many gaps will always be welcome.

ACKNOWLEDGEMENTS

It is a pleasure to record our gratitude to the hundreds of people, particularly bird ringers, who have made this work possible, by providing either new data on which our present knowledge of moult rests, or summaries of the result of their own research.

Particular thanks are due to Dr David Snow for allowing us to draw so heavily on his work "A Guide to Moult in British Birds". Our special thanks also go to Peter Marsack and David Partridge who undertook the tedious task of sorting the moult cards, preparing the data for final assessment and drafting the scatter diagrams; their help was made possible through a generous grant from the Worshipful Company of Grocers.

We are most grateful to the following people who reviewed parts of the text and made valuable suggestions and corrections and provided important unpublished material: M. Boddy, P. Ferns, M. Fletcher, D. M. Francis, R. Furness, H. Galbraith, D. Glue, M. P. Harris, A. J. Holcombe, C. Johnson, C. J. Mead, R. J. O'Connor, M. A. Ogilvie, D. J. Pearson, M. W. Pienkowski, I. M. Spence, R. Spencer, J. J. Tucker and A. B. Watson. We acknowledge that we alone are responsible for the remaining inaccuracies.

Mrs Elizabeth Murray painstakingly undertook the artwork. In the absence of both authors C. J. Mead very kindly undertook the task of seeing the work through the printers for which we are most grateful.

We wish to thank the following organisations for permission to use figures previously published elsewhere: British Ornithologists' Union, Cooper Ornithological Society, Deutsche Ornithologe-Gesellschaft and Institut Royal des Sciences Naturelles de Belgique.

PART I: INTRODUCTION

PLUMAGE AND FEATHERS

The ancestry of birds and the evolution of plumage.

The most obvious characteristics of birds, which readily distinguishes them from all other forms of life, is the plumage, the collective name for the outer covering of feathers; indeed a bird is defined taxonomically as a vertebrate with feathers. The fossil record strongly suggests that birds evolved from small reptiles which walked on their hind limbs but gives no clue to the means by which the feather evolved from a reptile-like scale: the earliest bird fossil *Archaeopteryx,* which dates back about 150 million years, already had feathers apparently indistinguisable in any important way from those of modern birds.

In the complete absence of a fossil record to help, the nature of the intermediate stages between the ancestral reptile and *Archaeopteryx* is still actively debated. The debate centres around whether down-feathers and warm-bloodedness (homoiothermy) evolved first with subsequent development of some of the feathers into quills permitting flight, or whether the ability to fly was the selective advantage giving rise to the evolution of some of the feathers into down for insulation.

The evolution of a gradual feather-by-feather replacement of plumage has been put forward as a difficulty in the development of birds, since reptiles usually replace the outer layers of skin and scales in one piece, or several large sheets. However, this problem was reduced with the discovery that the crocodiles, the nearest modern reptilian relative of the birds, replace the surface layers of their scales individually (Parkes 1966).

Whatever the evolutionary stages in the development of birds may have been the important point in the present context is that all fully grown birds have plumage which, in most cases, serves for flight as well as heat insulation, permitting the maintenance of a constant temperature, different from, and usually higher than, that of their surroundings. These two features, flight and homoiothermy, have been of paramount importance in enabling birds to exploit virtually the entire earth's surface to a remarkable degree.

The types of feather and their structure

The feather has, during evolution, become modified and diversified to a considerable extent and the basic forms are shown in Fig. 1.

1. Contour feather

The contour feather in all its various forms is one of the most familiar structures in nature. The majority of feathers are of this type, including the main flight quills of wing and tail (Fig. 1,A) and those forming the outer covering of the head and body (Fig. 1,B) — by definition all those forming the outlines or contour of the adult bird. The term 'contour' has often been used loosely to mean only the types of feathers which form the general covering of the head and body and the smaller feathers on the leading parts of the wings, as distinct from the 'pinions' of the wings and tail, and indeed the term was until fairly recently used in this way on the B.T.O.'s Moult Card, but has now been replaced by the term 'small feathers' (Fig. 7). Contour feathers are generally confined to distinct feather tracts (or *pterylae*) with naked or downy areas between (see page 10). In general flight feathers are asymetrical so that it is possible to tell whether an individual feather came from the right or left wing (or half-tail) whereas most other contour feathers are symetrical about the shaft.

The great variations in size, shape and colour of contour feathers are directly related to their function, for in addition to adaptions for flight and insulation, they also serve for buoyancy in aquatic species, streamlining in flight, weather-proofing,

Fig. 1

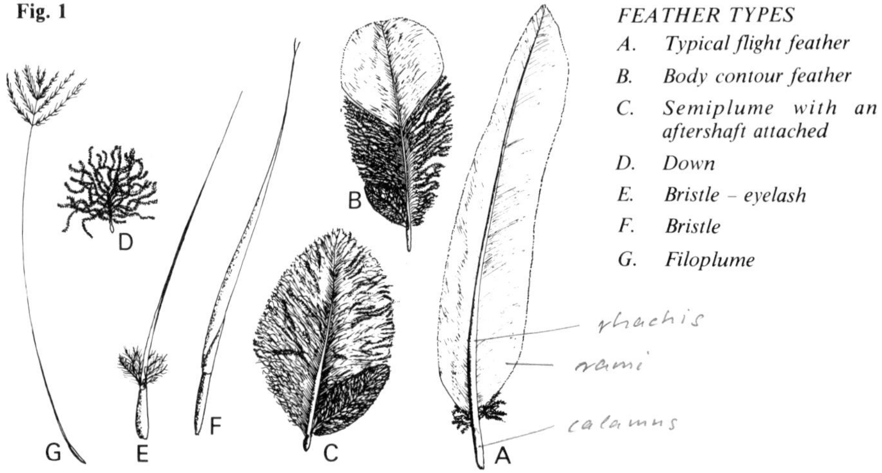

camouflage and, in the broadest sense, display.

The typical flight feather (Fig. 1,A) consists of a long, central tapering shaft whose short basal section *(calamus)* embedded in the feather socket *(follicle)* is hollow, cylindrical and devoid of side branches, and whose terminal section *(rachis)* is solid, angular and bears two rows of parallel side branches or barbs *(rami,* singular *ramus).* These barbs collectively form the web or vein of the feather. The calamus has two small openings, one at its tip and the other on its side at the point of junction with the rachis. Also at the junction of calamus and rachis in many veined feathers, is attached a second 'feather' called an aftershaft, usually smaller than the main feather, but which in some birds such as the Emu *Dromaius novaehollandiae* is of almost equal size. In fact the aftershaft is part of the complete feather in its primitive form, but has become much reduced or absent in most feathers during evolution, though it still occurs in some highly evolved groups of birds (see Miller, 1924).

The barbs (Fig. 2) are attached to the rachis at a slight angle, each inclined towards the tip of the feather. Each barb bears two rows of yet smaller side branches *(barbules),* slightly inclined towards the tip of the barb. The barbules are of two types depending on which side of the barb they are attached; those which point towards the tip of the feather *(distal barbules)* are flattened from side to side and have a series of tiny projections *(barbicels)* pointing downwards, those along the middle section of the barbule being hooked and called *hamuli.* The second type of barbules which point towards the base of the feather *(proximal barbules)* are similarly flattened but lack hamuli; instead they form the ridges into which the hamuli on the distal barbules hook so that collectively they form an interlocking lattice which is the essence of the feather vane's combined lightness and strength.

For a short length of the rachis at its junction with the calamus a number of the barbs have their barbules unlinked and downy *(plumulaceous).* The microscopic structure of these barbules varies considerably between species and groups and can be used to identify feather remains (Brom 1980, Chandler 1916, Day 1966). In other types of contour feather, for example those covering the general surface of the body, the overall structure is lighter and more loose-textured because there is a larger section of unlinked barbs.

2. Semiplume

The semiplume (Fig. 1,C) is rather like the contour feather of the general body

Fig. 2

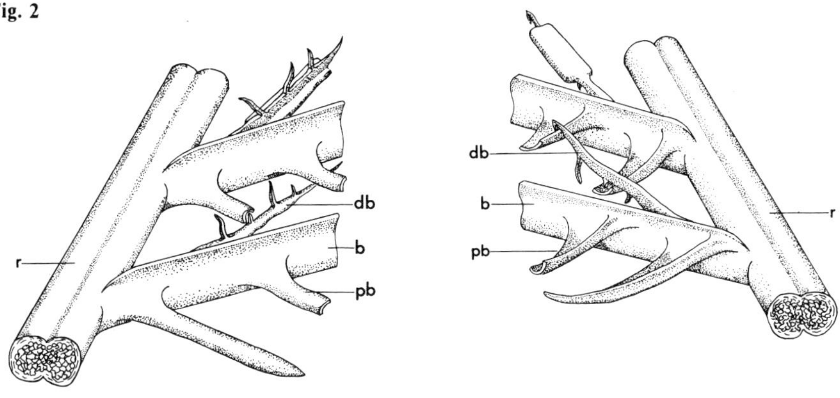

The fine structure of a feather. This enlarged diagrammatic view of part of the rachis (r) and two barbules (b) shows the same part from below (left) and above (right). The cut proximal barbules (pb) show their curled structures. The flattened top to the distal barbule (db) on the right, top is the site of iridescence seen on some feathers. The barbules are drawn much more widely spaced than they are in nature, for clarity. Drawings from 'Espinasse (1964), © British Ornithologist's Union.

covering, having a central shaft with lateral barbs bearing barbules but with the differences that it is entirely plumulaceous i.e. all the barbs are downy and lack hamuli so that they are free of attachment to each other and there is consequently no vane. Semiplumes occur typically at the edges of the pterylae and in the apteria (see p. 10) and are usually completely obscured by over-laying contour feathers. There is in fact a complete range of intermediate feathers between the semiplume and contour, but the fully developed semiplume is sufficiently distinct to justify the use of a separate term.

3. Down feather

Down feathers or plumules (Fig. 1,D) are, like semiplumes, usually concealed beneath the contours except in nestlings where they may make up the entire covering. They are entirely plumulaceous and usually have a very small rachis so that all of the barbs arise from approximately the same point at the end of the calamus, which is also very short. Down feathers if present at all are usually fairly widely distributed on the bird but in some species are restricted to the apteria. They are especially well developed in water birds.

4. Powder down

Powder down feathers are specially modified body feathers in which the calamus bears a number of long, soft filaments. In the most highly developed powder downs, the 'feather' grows continuously, breaking down at the tip to give the tiny, waxy, scale-like particles which make up the powder which gives a bloom to the plumage. Powder down feathers, which are present to varying extents in most birds, are usually grouped in paired tracts, often on the underparts, but sometimes they are scattered throughout the rest of the plumage. They reach their highest development in groups such as the herons (Ardeidae), tinamous (Tinamidae) and wood swallows (Artamidae). The resulting 'powder' is keratin and is generally regarded as being a water-proofing dressing for the plumage (Lucas and Stettenheim 1972).

5. Bristle

Bristles (Fig. 1,E,F) are extremely modified and specialised contour feathers. In form they vary from a simple shaft devoid of side branches to a long central shaft devoid of barbs except for a loose-textured cluster at the base. Bristles are usually found over the nostrils and around the gape but in some species are also found around the eye (Zonfrillo 1982). In some birds from quite diverse order e.g. Ostrich *Struthio camelus,* some hummingbirds (Trochilidae) and a number of cuckoos (Cuculidae) — bristles around the eyes form 'eyelashes' very like those found in mammals. The functions of bristles have received little critical study.

6. Filoplume

Filoplumes, like bristles, are extremely specialised and modified feathers which resemble mammalian hairs. Always found in direct association with contour feathers, they are of two types. The first type (Fig. 1,G) which generally has a small tuft of barbs at the tip and which are usually completely obscured by the neighbouring feathers and occur singly or in a small cluster at their bases; they appear to arise from the same feather papilla. The second type lacks barbs and is frequently elongated sufficiently to protrude beyond the contour feathers. When, as is often the case, such filoplumes occur on the nape and upper back they form a distinct 'hairy' mane; an extreme form of this condition is found in the Hairy-backed Bulbul *Hypsipetes criniger* of S.E. Asia.

7. Other feather types.

Many other modifications to the basic feather structure occur, some of which do not fit easily into the categories described and many forms and intermediate between them.

The development of the feather

Each feather is formed from a tiny cone-shaped structure *(papilla),* richly supplied with blood deep in a pit *(follicle)* in the skin. The young feather is enclosed in a flexible waxy sheath and grows below to force out the old feather and emerge from the follicle, still enclosed in its protective sheath. At this stage it is described as 'in pin'. When the tip of the feather is well clear of the follicle it ruptures the end of the sheath and the barbs previously compressed into a cylindrical shape, spread out to form the vane. The sheath remains as a collar round the calamus of the feather (the feather being described as 'in sheath') until it is fully grown when it flakes away. It is of interest to note that the nestlings of some species such as the Kingfisher *Alcedo atthis* have the entire length of their feathers ensheathed until virtually fully grown, the sheathing dropping off only shortly prior to fledging; this is presumably an adaptation for the protection of the new feathers from the filth which abounds in the nesting hole.

For a detailed description of feather development see 'Espinasse (1964) and Stettenheim (1972).

The arrangement of plumage (pterylosis) in adult birds

Feather tracts (pterylae)

Birds have their plumage distributed in distinct bands or tracts *(pterylae,* singular *pteryla)* with areas between them *(apteria,* singular *apterium),* which may be either completely devoid of feathers, sparingly covered with semiplumes or densely covered in down feathers. Externally the apteria are usually completely obscured by the overlapping contour feathers in the adjacent tracts. Certain groups of birds have the feathers more or less evenly distributed over the body surface; examples are the penguins (Sphenisciformes), the screamers (Anhimidae) of South America, the mousebirds (Coliiformes) of Africa and the Ostrich *Struthio camelus* and its relatives. Clench (1970) notes that even in these forms there is always at least one apterium. The arrangement of tracts varies considerably from one order of birds to another and has been used as a taxonomic character to suggest evolutionary relationships (e.g. Compton 1938). Many species are illustrated in Nitzsch's classic work (1867), and references to other works are given in Part II.

Typically seven or eight tracts are recognised, the exact number depending on the author's opinion (Fig. 3).

Fig. 3

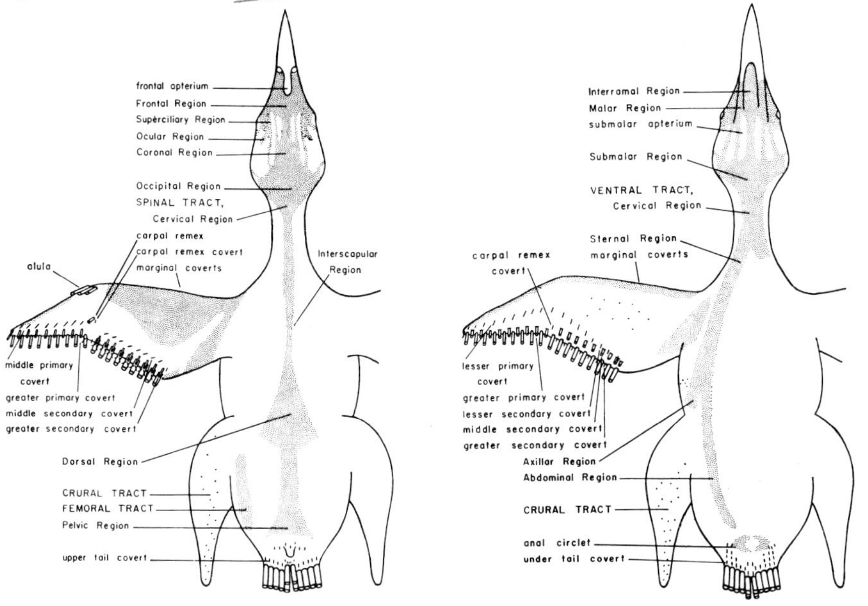

The distribution of feather tracts on an American Water Pipit Anthus spinoletta alticola: *the dorsal view is on the left and ventral on the right. Taken from Verbeek (1973) © Cooper Ornithological Society.*

1) the *capital* tract covering the crown from the base of the bill to the back of the skull.

2) the *spinal* (or *dorsal)* tract running along the back from the neck as far as, but excluding the tail.

3) the *caudal* tract made up of the rectrices (tail feathers), the upper and under tail coverts.

4) the *ventral* tract, which is very variable and divided centrally for at least part of its length, running from the throat to the vent.

5) the *humeral* tract covering the leading edge of the wing and shoulder region.

6) the *alar* tract made up of the remiges and their coverts.

7) the *femoral* tract running diagonally across the thigh.

8) the *crural* tract crossing the remainder of the leg.

Some authors regard 7 and 8 as subdivisions of the *hind-limb* tract, and 5 and 6 as subdivisions of the *wing* tract.

Remiges (Fig. 4)

1. Primaries

The primaries are those flight feathers on the outer part of the wing attached to the part of the skeleton homologous to the human hand. In flying birds, the primaries may number from nine to eleven depending on the species. Some species have an extremely small 'extra' feather outside the tenth primary, referred to as the *'remicle'*, and opinions differ as to whether this is a true primary or a covert feather (Stegmann 1962, Stresemann 1963b, Stresemann and Stephan 1968). The number of primaries does not usually vary within a species but in those with a very reduced outer primary e.g. waders of the family Scolopacidae, this may occasionally grow to a size similar to that of the other primaries, and there are records of birds with, for example, only eight instead of nine primaries, but such anomalies are very rare.

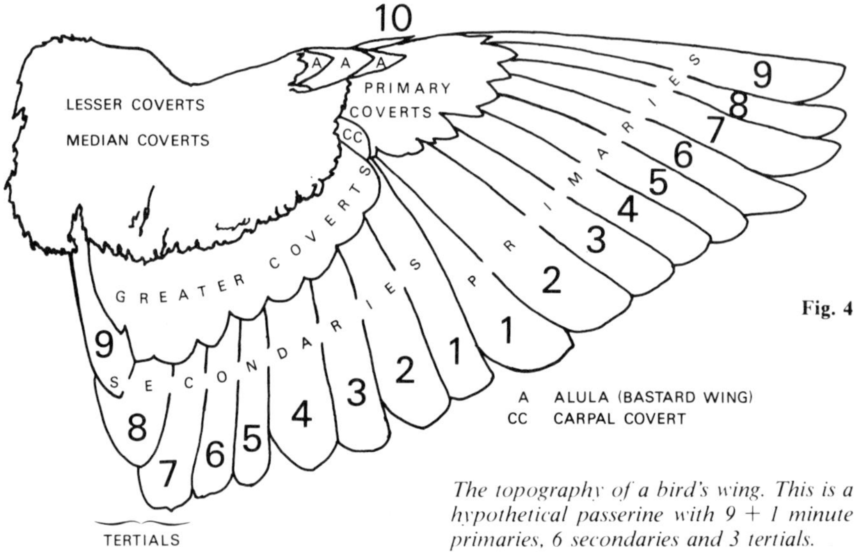

Fig. 4

The topography of a bird's wing. This is a hypothetical passerine with 9 + 1 minute primaries, 6 secondaries and 3 tertials.

Primaries may be numbered descendantly, that is from the carpal joint (wrist) outwards from the body, or ascendantly from the outside towards the carpal joint. In this work the primaries are numbered descendantly. This has the advantage that P1* in any species is homologous to P1 in every other species of bird, as is P2 etc. Furthermore the descendant system is preferred since if primaries are 'lost' during the course of evolution it is usually from the outside of the wing, and in the great majority of species moult progresses descendantly through the primaries. The ascendant system of numbering is widely and traditionally used in Europe (e.g. Hartert 1910-1922, Vaurie, 1959, Svensson 1975) and was used in the Handbook of British Birds (Witherby et al. 1938-1941). The Handbook of the Birds of Europe, the Middle East and North Africa (Cramp and Simmons 1977, 1980, et seq.) numbers the primaries descendantly. When referring to the literature it is important to note which method of numbering is used (Ashmole et al. 1962).

Some flightless species have an increased number of primaries: for example the Ostrich has 16 which seem to be used for display, whereas others such as the kiwis (Apterygiformes) have only four.

*Explanation of abbreviations on page 35.

2. Secondaries

The secondaries are those feathers attached to the ulna (the equivalent of the human forearm). They are always numbered ascendantly from the carpal joint towards the body. The secondaries, unlike the primaries, vary enormously in number between species. The number depends greatly on the configuration of the wing, on the size of the bird, and particularly on the length of the forearm. The longer the wing, the more secondaries are developed at the inner end of the tract. The hummingbirds have six or seven; most passerines usually have nine full-length ones though some species may have as many as 14; the woodpeckers (Picidae) have eleven; the New World vultures (Cathartidae) have 18 to 25 and the albatrosses (Diomedeidae) carry up to 40.

Each primary and secondary has a small covert feather covering its base. These are termed *tectrices* (singular *tectrix*) and comprise the greater upper primary coverts and the greater upper secondary coverts. These are commonly referred to as the primary coverts and the greater coverts respectively.

3. Alula or bastard wing

On the upper leading edge of the wing, just distal to the carpal joint is the alula or bastard wing, a group of from two to seven but most often three or four small feathers attached to the pollex (equivalent to the human thumb). They are usually graded in size, the outermost being the largest.

4. Carpal remex

On the upper wing surface, between the innermost primary and outermost secondary is often found an extra remex, the carpal remex, with its own covert, the carpal covert. In some Galliformes and in gulls (Laridae) the carpal remex is approximately half as long as the secondaries (to which series it probably belongs) but in other groups it may be very small and lacking a carpal covert (e.g. woodpeckers); in yet other groups the carpal covert is present without the carpal remex (e.g. most Passeriformes).

5. Tertials

In many passerine families the three innermost secondaries (numbered seven to nine) are often differentiated by shape or colour or by moult sequence from the rest of the series. These are frequently termed the *'tertials'*.

6. Underwing coverts

The greater and median underwing coverts are 'reversed' their concave surface faces away from the wing to produce a more efficient aerofoil shape.

Rectrices

The rectrices or tail feathers vary greatly in number from four to more than 30. They usually occur in even numbers, conventionally being numbered in pairs from the centre outwards (centrifugally), but occasionally an individual bird may be found with an uneven number of rectrices (Hanmer 1981). The most frequent number of rectrices is twelve (six pairs), but some notable exceptions include the divers (Gaviidae) 18-20, waterfowl (Anseriformes) 14-24 and Swift *Apus apus* 10.

The shape and length of the tail varies enormously in a way generally related to mode of flight or display, e.g. broad and rounded in soaring species; deeply forked and often with the outermost feathers forming streamers in strong fliers like terns (Sternidae) and swallows (Hirundinidae); modified to produce a 'drumming' sound during display flight in some snipes *(Gallinago* spp.*)*, or elaborately developed to produce a visual display as in the lyrebirds *(Menura* spp.*)* of Australia. The rectrices of grebes are vestigial.

The total number of feather of birds

The total number of 'contour' feathers of all types on a bird is approximately related to its size. The larger the bird the greater the number of feathers present: some hummingbirds may have less than 100, the Swallow *Hirundo rustica* about 1,500, while the Whistling Swan *Cygnus columbianus* may have over 25,000. However the density of feather follicles also varies, and in the Gentoo Penguin *Pygoscelis papua* 46 per square centimetre (at various stages of development) have been recorded. The number of feathers on any individual may vary seasonally. Thus Brooks (1968) found that the plumage of redpolls *Carduelis flammea/hornemanni* in Alaska was 31% heavier in November than in July. Such variation probably has survival value in areas with marked seasonal temperature fluctuations, greater insulation being provided by more feathers in the colder season. Wetmore (1936) details the number of feathers for various North American species.

The colour of birds

White light consists of the full range of colours of the spectrum. The colour of an object depends on which of these colours are reflected and which absorbed by the object. The quality of the colour will also be affected by the amount of scattering of various colours. There are two types of colour in birds, the *pigmentary colours* and the *structural colours*. The first (termed 'biochromes') depend on the presence of pigment in the form of granules or globules in the feathers; the second (termed 'schemochromes') on the microscopic structure of the feather.

Pigmentary colours

These occur in the bill and exposed skin of birds as well as in the plumage. The pigments are of three types — *melanins*, *carotenoids* and *porphyrins*. Melanin is found as distinct rather insoluble granules, which may be black, brown, reddish-brown or even pale yellow, and are produced by specialised cells in the developing feather. The black and dark brown forms are usually termed *eumelanin*, the light-brown and yellow forms *phaeomelanin*. Both types of melanin may be present in the same feather but usually the eumelanins are concentrated towards the tip, the phaeomelanins at the base.

Unlike the melanins, the carotenoids are fat-soluble and are generally responsible for red, orange, yellow and violet colours. Carotenoid pigments are not synthesised by birds and must be obtained in the diet; they may subsequently be metabolised to other pigments or, as in the case of certain flamingos (Phoenicopteridae), used directly. If a carotenoid pigment is absent from the diet during feather growth, it will also be lacking in the plumage. However the presence of such pigments in the diet may also result in unusually coloured birds such as 'orange' Greenfinches *Carduelis chloris* and Linnets *Carduelis cannabina*. Instances of such pigmentation are generally uncommon in the wild (Washington and Harrison 1969) but form the basis for many of the more colourful captive finches.

Porphyrins, which are also fat-soluble, occur in the down of some birds and also in the contour feathers of owls (Strigidae) and result in red, green and brown colours. Porphyrins are often particularly light sensitive and may fade rapidly in sunlight.

Some plumage colouration may be the result of a combination of pigments or even of various pigments and structural colours. Green colour may be produced in this way with a yellow pigment in the outer layers of the feather, which filters out some of the light, overlying a layer rich in air spaces which scatter more of the light, leaving wavelengths which in combination appear green.

Structural colours

These are of two types, the iridiscent and non-iridiscent. Iridiscent colours change with the angle of view. The structure which produces the iridiscence is in the barbules, which are flattened for part of their length and twisted (Fig. 2). Since this modification is only possible by sacrificing the mechanical strength of the feather, it is generally absent from flight feathers. The structural modification usually has a background of melanin granules arranged in a single layer just below the surface of the barbules and sometimes the granules themselves produce the irridiscence e.g. in some hummingbirds.

Non-iridiscent structural colours remain unchanged no matter how the angle of view is changed. In some feathers, it is not pigment particles which reflect light but minute air-pockets. Feathers with this type of structure are not blue when viewed with transmitted light (i.e. held up to the light) but brown due to the melanin granule. White feathers have a microscopic structure which scatters incident light in all directions from myriads of particles with different refractive properties.

Cosmetic colouration

An additional source of colour in birds is the oily secretion of the uropygial (preen) gland. This apparently contains carotenoid pigments in some species such as the Great Pied Hornbill *Buceros bicornis* which are transferred from the gland to the white parts of the plumage. Because they are unstable pigments which bleach in sunlight, they have to be renewed frequently (Vevers 1964). Stegmann (1965) suggested that the uropygial secretion may be the source of the pinkish suffusion on the white feathers of the Black-headed Gull *Larus ridibundus* in spring. The phenomenon may be quite widespread and deserves further study.

Other applied discolouration of the plumage may occur as in the reddish 'iron' staining of swan plumage, and 'industrial' melanism with soot and other dirt in urban areas (Harrison 1963).

For further information on the colouration of feathers see Cohen (1966), Lucas and Stettenheim (1972), Stettenheim (1972) and Rawles (1960).

Functions of plumage

Streamlining and insulation

The feathers are arranged over the general surface of the bird so that they streamline the body and wing surface during flight; they point backwards towards the tail and each feather is overlapped by the one in front to present a barrier to wind and rain, functioning in a similar way to the tiles on a roof. A layer of air is thus trapped around the body, insulating it from the outside. The insulation is increased by the down feathers next to the skin and by the downy bases of the contour feathers.

Feathers and flight

A full description of the functional anatomy of wings and the principles of flight are not appropriate here and reference should be made to Pennycuick (1972, 1975). It is sufficient to emphasise the crucial importance of the wing and tail feathers, particularly the primaries, for flight and the need both to maintain them in good condition and, if flight is to be attempted during the moult period, to replace them in a gradual sequence. The contribution of the wing coverts and inner secondaries in protecting the other flight feathers when they are at rest should not be overlooked.

Waterproofing and buoyancy

In water birds, the plumage forms a water repellent layer; the body plumage is generally denser, and is more oily than in other birds thus excluding water by surface tension, and there is copious down next to the skin. The air trapped in the plumage increases bouyancy and insulation. Before diving some species (e.g. Great Crested Grebe, *Podiceps cristatus)* flatten the plumage closer to the body by muscular action to expel some of the air to facilitate submersion.

Display and camouflage

Since the plumage forms the outer covering of the body it is the most obvious visible part of the bird. During the course of evolution the structure, colouring and patterning of feathers has become modified for use in various activities — for example sexual, aggressive and threatening displays and for hiding from predators. In many species cryptic colouration, evolved for this last purpose, may at rest conceal patches of bright colour or contrasting patterns used for signalling or display. Many species have white rump patches or wing bars invisible when the bird is at rest with its wings folded.

Feather maintenance

Feathers, being complex and vital structures, must be cared for to prolong their life as much as possible. Preening with the bill serves to rearrange the feathers, relocate barbs which have become unlocked, remove feather-lice (Mallophaga) which feed on the feather protein, and also to spread oil from the uropygial gland over the surface of the plumage to assist in 'weather proofing'. Various other activities such as dust-, sun-, and smoke-bathing and "anting" have been described and are often considered to relate to feather maintenance (see Simmons 1964, Stainton 1982).

Abrasion

Feathers are generally extremely strong, light and durable but in spite of scrupluous and regular maintenance, they are subject to progressive wear and tear. The vanes of the feathers gradually abrade away at the edges and tips, and the barbules and barbicels which hold the barbs of the vane together, deteriorate in response to the enormous forces exerted on them during flight, to the friction with neighbouring feathers, and rubbing against things such as branches or the ground. Sunlight, in particular, acts adversely on the feather protein and its associated pigments giving rise to a breakdown of physical structure and colour. Feather-lice also take their toll. Eventually all of the functions of the feathers are impaired. Pigmented feathers generally abrade less quickly than white ones and this can be clearly seen in pied feathers e.g. the white 'mirrors' on the outer primaries of some gulls are often severely abraded by late summer whereas the surrounding black area is only slightly abraded (Averill 1923).

In many birds the breeding plumage is revealed by the abrasion of feather tips; for example the black heads of the male Brambling *Fringilla montifringilla* and Reed Bunting *Emberiza schoeniclus* are almost completely obscured by light feather tips when the plumage is fresh after a moult, the tips wear away late in the winter for the black to show through in the spring. In the Bee-eater *Merops apiaster* the fresh primaries are green, these then 'weather' to bright metallic blue.

The need to moult

In view of the inevitable abrasion, if a bird is to survive the plumage must be replaced before the impairment of function is sufficient to handicap the bird significantly. The overall process of replacement of the plumage is called the moult*; the term generally being used to include both the loss of the old feathers and the growth of the new. If a bird loses feathers accidentally during the year these usually will be replaced but this is not a result of moult. In moult the old feathers are actually pushed out by the new ones growing in the follicle below (G. E. Watson 1973). Moults fall into two broad categories: complete moult in which all of the feathers are replaced (usually once a year), and partial moult in which only some of the feathers are replaced.

*'molt' in current North American usage.

MOULT AND THE ANNUAL CYCLE

THE COMPLETE MOULT OF ADULTS

Birds display a great variety of patterns in their moult, both in relation to the timing of the moult in the annual cycle, and in the sequence in which the feathers are replaced (Palmer 1972, Payne 1972, Stresemann and Stresemann 1966). Our knowledge and understanding of these variations is still very far from complete.

In most bird species, especially in temperate regions, moult, migration and breeding do not overlap significantly, and moult frequently occurs soon after breeding, especially in passerines (Fig. 5).

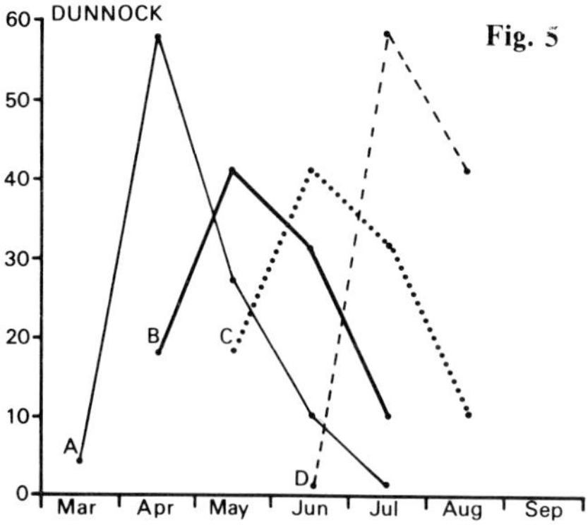

Fig. 5

The timing of breeding and moult in Dunnocks Prunella modularis *is shown:*

A First egg date from national records

B First egg date from detailed study in Oxford, 1960

C Theoretical fledging date from B

D Percentage of Dunnocks starting to moult in each month: from Ginn (1975)

However some species do not have a clear-cut segregation of these activities and do overlap breeding and moult to some extent, this being more noticeable in the tropics and in areas with irregular rainfall where there is opportunistic breeding (e.g. Britton 1978, Diamond 1974, Foster 1975, Gaston 1981, Keast 1968, Payne 1969).

It has frequently been suggested that the temporal separation of these activities in the annual cycle indicates that they are to some extent in competition for one or more resources; food has been regarded as being especially important. Feathers account for a considerable proportion of the weight of a bird (Turček 1966). For example in the Bullfinch *Pyrrhula pyrrhula* feathers make up c10% of the total live weight and c40% of the total dry weight (Newton 1966). It is considered that a bird rearing young already has a high food requirement and generally cannot find enough additional food for its own moult. Further a bird is unlikely to undertake a long distance migratory flight without a full complement of flight feathers.

The process of moulting places additional energy demands on a bird because:-

i. energy is required to produce replacement plumage.
ii. extra energy is required to regulate body temperature while feather insulation is reduced.
iii. extra energy is required when gaps in the wing caused by dropped/growing feathers result in less efficient flight.

In addition to 'energy', birds require protein for feather production and certain amino-acids such as cystine and cysteine which they are probably unable to synthesize in any quantity, and which therefore must be taken in in the diet. It has not been possible to measure these various factors separately but a number of attempts have been made to estimate energy requirements. Dolnik and Gavrilov (1979) found that the energy demand from different factors varies throughout the moult, a larger proportion of productive energy being used for feather formation in the early stages of moult than later.

The situation is clearly complex and much work remains to be done. However the present trend of thought is that, for those species which spread the moult over a considerable period of time, it is probable that the process does not exert a major physiological stress on the bird, the increased energy requirements per day being relatively low (e.g. Kendeigh *et al.*

1977, King 1974, 1980, Payne 1972). Even in species such as the Lesser Snow Goose *Anser caerulescens* and the Barnacle Goose *Branta leucopsis* where moult is rapid, it appears that increased nutrient demand can be met through increased food intake (Ankney 1979, Owen and Ogilvie 1979). Owen and Ogilvie (1979) have suggested that the northward moult-migration undertaken by many geese enables them to utilise vegetation at the start of the growing season when it is more nutritious and more easily digested. The penguins (Sphenisciformes) are exceptional in that the complete moult occurs whilst the birds are incubating during which they undergo a fast (Croxall 1982).

The hormonal control of moulting and the effects of varying temperature and day length (photoperiod) also remain inadequately studied and work to date suggests that there are various mechanisms at work in different species. Recent reviews are given by Murton and Westwood (1977) and Meier and Ferrell (1978).

Thus it is still not possible to explain the complex relationship between moult and other activities in the bird's annual cycle or how these are maintained and controlled.

Moult Strategies

There are several different moult strategies which birds may employ to fit the annual complete moult into their annual cycle. These are most constrained for long-distance migrants and six have been listed by Stresemann and Stresemann (1966) as follows:

a. Suspended moult.
b. Moult after autumn migration.
c. Moult before autumn migration.
d. Moult during the breeding cycle.
e. Moult both before and after autumn migration (two complete moults in each year).
f. Periodic stepwise moult *(Staffelmauser)*

These strategies are also employed by resident species which will, because they do not have to migrate, have more time available during the autumn (or after breeding for species in aseasonal habitats) to moult. Some examples for species from arctic, temperate and tropical areas are given below.

The arctic region

In summer food is often abundant and daylight hours available for feeding are plentiful. However this active season is short and the competition great. Thus often there is barely time for a bird to establish territory, breed and moult before the cold weather sets in again. There is a range of strategies which birds can employ dependant on the latitude and species, (Salomonsen 1972). Some manage a complete breeding cycle followed by a complete moult before winter sets in or migration must be undertaken. Usually this is only possible for birds starting moult while still attending young or moulting very rapidly, some may sometimes become flightless for a brief period. Often both features are found as in Snow Buntings *Plectrophenax nivalis* and Lapland Buntings *Calcarius lapponicus* in Greenland, and various other passerines in Scandinavia (Haukioja 1971a, 1971b).

The Glaucous Gull *Larus hyperboreus* breeding in Iceland and Alaska starts moulting during the incubation period, while the Ivory Gull *Pagophila eburnea* begins moulting in March or April and has renewed many, if not all, of its primaries by the time that egg-laying begins towards the end of June or the beginning of July (Stresemann 1963c).

Other species have evolved other ways of dealing with the situation and may not undergo a complete moult immediately after breeding (e.g. some divers, the Puffin *Fratercula arctica*). Others may start and then suspend* the process; moult may then be resumed and completed in a favourable area en route for winter quarters or on arrival in the winter quarters. There are many examples of this strategy in the waders, for example Grey Plover *Pluvialis squatarola*, Dunlin *Calidris alpina* and Wood Sandpiper *Tringa glareola*.

The temperate region

The situation where a lengthy moult 'projects' backwards into the breeding

*see Glossary, page 101.

period is not uncommon in non-passerine species even in the temperate zone. However in passerines it is much less frequent and is generally found in migrants sandwiching moult between a late brood and imminent migration. For sedentary species there is often much more time to tend the last brood of young and to moult before winter sets in, when the food supply and the daylight hours available for exploiting it diminish. There is some evidence to suggest that in certain species adults may start to moult the primaries after rearing a first brood, or, more likely, after a failed first nesting attempt, and then suspend moult during the second/replacement attempt. This has been recorded in the Willow Warbler *Phylloscopus trochilus* and Whitethroat *Sylvia communis* (R. A. Cawthorne pers. comm.) and may be more widespread than the few records suggest, being more likely to occur in those species which drop the inner primaries shortly before the young fledge or become independent.

Migrants which breed in the temperate region but winter further south may undergo a complete post-breeding moult before the autumn migration (e.g. Redstart *Phoenicurus phoenicurus*), may start moult on the breeding ground, suspend and finish in the winter quarters (e.g. Turtle Dove *Streptopelia turtur*) (Mead and Watmough 1976, Swann and Baillie 1979), or may undergo a complete moult in the wintering area (e.g. Reed Warbler *Acrocephalus scirpaceus*).

The strategy adopted appears to be generally related to the location of the breeding and wintering areas of the species concerned and the distance between them. For example, the majority of warblers of the genus *Sylvia* moult before reaching wintering areas, usually on or near the breeding grounds but sometimes during breaks in migration. This is probably related to the fact that they perform relatively short migrations mostly to areas not far north of the equator or at most, as far as 20°S; the Garden Warbler *Sylvia borin* and the eastern race of the Common Whitethroat *Sylvia communis icterops* whose wintering ranges extend south to 30°S and about 20°S respectively, generally do not moult until reaching Africa. Similarly, among species or genera which undergo a complete moult during their stay in winter quarters, those species or populations within species which move furthest south tend to delay the moult later. This situation is found for example in the genera *Locustella, Acrocephalus* and *Hippolais*, as well as in some shrikes of the genus *Lanius* (Pearson 1973a).

Examples of intraspecific variation are provided by the Great Reed Warbler *Acrocephalus arundinaceus* and the Sedge Warbler *A. schoenobaenus,* most of which renew their plumage soon after arriving in Africa, while a minority that winter further south moult just prior to the northward migration. An interspecific comparison is provided by the Reed Warbler and the Marsh Warbler *Acrocephalus palustris.* The former, which winters mainly in East Africa between 10°N and 15°S, moults chiefly in late autumn and early winter, while the latter, wintering from about 5°N to 30°S moults in February and March immediately before departure for Europe. Observations from various parts of Africa suggest that some species exploit temporarily abundant food sources for moulting at some point on their way south, arriving at the southern limits of their winter range in new plumage (Pearson 1973a).

All these species follow the normal rule which is to have a single, annual complete moult but one exceptional species is the Willow Warbler which has two complete moults per year, one in the breeding area after breeding the other in the winter quarters. The closely related Chiffchaff *Phylloscopus collybita* has only one complete moult after breeding and before migration. The difference is probably related to the average distance of migration since the Willow Warbler generally breeds further north and winters further south than the Chiffchaff, and its plumage may not be sufficiently durable to las for two full-length migrations.

Pearson (1973a) notes that long distance migrants "gain two obvious advantages from delaying wing moult until arrival in the tropics. Firstly, their flight feathers are in a relatively fresh condition for the spring migration, which tends to be more rapid and demanding than the autumn journey. Secondly, there is more time available for moult in winter quarters than in the few weeks between completion of breeding and commencement of autumn migration". Alerstam and Högstedt (1982) have speculated that moult may be delayed in those species which defend territories in the winter quarters, and which thus need to migrate as soon as possible after breeding in order to establish a territory. An example is provided by the Greenish Warbler *Phylloscopus trochiloides,* which defends a territory in winter quarters and delays the complete moult until the spring (Price 1981).

The tropics

In the tropics the average temperature is high and the daylength more or less uniform throughout the year, allowing diurnal feeders a uniform feeding time each day. There are, however, other types of 'seasonality' based on wet and dry seasons, on flowering and fruiting seasons and, in the oceans, on seasonal variation in temperature, currents and upwellings. Hence, although there are periods of food abundance there may also be much competition, high predation and distinct periods of low food supply. It is usual therefore to find that moult still occurs outside the breeding season, but in a number of species the two activities regularly overlap (e.g. Ward 1969). Also, due to a prolonged period of parental care in the post-fledging period of some species, adults may start wing moult whilst still feeding young (Fogden 1972). Extensive migration tends to be the exception hence there is not the problem of fitting a moult between breeding and migrating.

In certain seabirds nesting in tropical oceans, the annual cycle is determined very directly by the need to moult and breed, the two activities may alternate on a less than annual cycle in some species. The Sooty Tern *Sterna fuscata* and Black Noddy *Anous tenuirostris* are examples where the non-breeding part of the year is occupied with the moult. The Sooty Tern on Ascension Island breeds at intervals of 41 to 51 weeks and moult is completed before incubation begins but the new moult cycle may start before the chicks are fully grown. By comparison the Black Noddy on Ascension may occasionally start breeding well before the end of the moult, but generally speaking the two activities are mutually exclusive (Ashmole 1962, 1963). For further details of the enormous variety of regimes in seabirds and especially those in the tropics see Diamond (1976) and the extensive review by Ashmole (1971).

SEXUAL DIFFERENCES IN THE TIMING OF MOULT

There is frequently a difference between the sexes in the timing and rate of the post-breeding moult; this may result from sexual differences in parental involvement in breeding activities. For example the female Reed Bunting *Emberiza schoeniclus,* continues parental duties longer and moults later than the male; apparently forcing a more rapid moult to permit completion before the winter. The situation appears to be even more marked in the Pied Flycatcher *Ficedula hypoleuca* studied in Saxony where the earliest males began to moult during the first part of June but the females, taking a greater and greater share of the parental duties as the males lose interest, delay the start of moult until the end of July or early August, just before the southward autumn migration.

In the Mallard *Anas platyrhynchos,* as in many other northern hemisphere ducks, the drake moults its wing feathers about June after it has undergone a body moult into 'eclipse' plumage, a well camouflaged attire not unlike that of the female, and undergoes such a rapid moult that it becomes flightless for about four weeks. The female by contrast delays the moult until after the young have become independant in July or August. In breeding Mute Swans *Cygnus olor* this is reversed: the female drops her remiges two to three weeks after the cygnets hatch but the male does not do so until the female has regained the power of flight some weeks later; in moulting flocks of non-breeders and failed-breeders, on the other hand, males moult before females. (Mathiasson, 1973).

Sexual differences are also found in some of the smaller raptors. In the Sparrowhawk *Accipiter nisus* the male remains very active, bringing food to the incubating female which starts to moult at about the time of egg laying; the male however does not start moulting until about 26 days later. Both males and females may suspend moult during the period of peak food demand by the young at about the time of fledging. Similarly, the male Kestrel *Falco tinnunculus* starts to moult about 14 days after the female.

Perhaps the most extraordinary and flexible adaptation of moult is found in certain hornbills (Bucerotidae). The female is sealed in the nest-hole throughout much of the nesting period, being fed by the male through a narrow entrance hole. During her confinement, she does not need to fly and is well protected from predation and so is able to undergo a very rapid moult of the wing and tail feathers. The male himself undergoes only a gradual moult. In years when a brood is not reared the female moults in the same way as the male (Moreau 1937, Stresemann and Stresemann 1966).

PRE-BREEDING MOULT

In some species, and particularly those with distinct breeding and non-breeding plumages, there is a partial moult of some of the head and/or body plumage. For example the Ptarmigan *Lagopus mutus*, has an all-white plumage during the winter when there is snow on the ground, and a brownish plumage during the breeding period, so it is well camouflaged throughout the year; the Black-headed Gull *Larus ridibundus* replaces the white winter head with a dark chocolate brown mask by a partial moult in time for the spring courtship period; the various species of waders with a distinct breeding plumage e.g. Knot *Calidris canutus* have a total moult of the body feathers and often some inner wing feathers prior to the breeding season.

Spring moult involving body plumage and sometimes inner secondaries also occurs in some passerines, such as the Pied Flycatcher and some of the Motacillidae.

THE EXTENT AND TIMING OF THE MOULT OF JUVENILES

The body feathers of juveniles are usually weaker and looser in texture than those of adults and in temperate regions juvenile birds usually undergo a partial moult of the entire body and head plumage and of the smaller wing-coverts at about the same time that the adults are undergoing their complete moult. A complete moult then occurs in the second year of life. Often some, or all, of the greater coverts are also renewed and less frequently the tertials and innermost secondaries and some or all of the rectrices. Post-juvenile moult provides a useful method of separating the age classes in otherwise difficult-to-age species as there is often a difference in colour, pattern or shape between juvenile and replacement 'adult' feathers. In some species the moult produces a complete transformation, such as the Robin *Erithacus rubecula* from a dappled spotty-breasted juvenile to a fairly uniform brown, organge-breasted 'adult' in one step. In others, such as certain gulls (*Larus* spp.) and gannets (Sulidae), the speckled brown and white juvenile plumage change progressively, over several annual moults to the mainly white or grey and white adult plumage.

Certain species are atypical in that the juveniles undergo a complete moult of all the plumage soon after fledging. British species which have a complete post-juvenile moult are: Shore Lark *Eremophila alpestris*, Skylark *Alauda arvensis*, Woodlark *Lullula arborea*, House and Tree Sparrows *Passer domesticus* and *P. montanus*, Starling *Sturnus vulgaris*, Bearded Tit *Panurus biarmicus*, Long-tailed Tit *Aegithalos caudatus*, and Corn Bunting *Miliaria calandra*. Occasionally some juvenile finches (Fringillidae) undergo a complete moult during their first autumn, this being more frequent at southerly latitudes.

The possible causes of a complete post-juvenile moult in such a limited selection of birds have been surprisingly neglected. C. J.

Mead (unpublished) has noted that, for British species, a complete post-juvenile moult occurs in those in which the young are fed on a largely vegetable diet (e.g. larks, sparrows, Corn Bunting). He suggests that such a diet may be deficient in amino-acids essential for the development of strong plumage. The Starling, on the other hand, feeds its young on an invertebrate diet. Tinbergen (1981) found that in an experimental situation in Holland Starlings reacted to an increase in brood size by changing their foraging tactics from seeking desirable, but more difficult to obtain, caterpillars, to less desirable, but more easily obtained, leatherjackets (Tipulidae). Young Starlings usually produce faecal sacs which are removed by the adults but those fed with a high proportion of leatherjackets tend to foul both the nest and themselves with wet faeces. Such fouling apparently increases heat loss through reduced insulation of both the nest and the plumage (Tinbergen 1981, Westerterp *et al.* 1982), and also is likely to have a detrimental effect on the feathers themselves. Leatherjackets form a major part of the diet of Starlings in Scotland (Dunnet 1955). If selection pressures on Starlings are such that they need to rear larger rather than smaller broods and if this can only be achieved by increasing the proportion of leatherjackets in the diet with a concomittant deterioration of plumage quality, then it would presumably be advantageous to have a complete post-juvenile moult. Bearded Tits are exceptions, being temperate zone representatives of the tropical family Timaliidae, and Long-tailed Tits are closely related. It is noteworthy that in Sarawak Fogden (1972) recorded a complete post-juvenile moult in all birds studied, both passerine and non-passerine. He suggested that juvenile plumage may be of poorer quality than that of adults and noted that individual remiges and rectrices of juveniles may be up to 30% lighter than the corresponding feathers in adults. Clearly the topic deserves further study.

In many game birds (Gallifomes and Turnicidae) there is an interesting adaptation, for the chicks can fly when very young, when they are only a fraction of the adult size. The flight feathers grown early in development are shorter than those required for flight by the fully grown bird, but the later grown juvenile feathers are more in proportion to the larger bird. As a result the innermost primaries are moulted and replaced by longer feathers before the outermost ones are fully grown. In *Turnix* spp. (bustard quails) this situation is taken to an extreme as the second generation innermost primary (P1) forces its predecessor out as soon as it has completed growth at 12 to 15 days old. In the tropics, the replacement of the primaries is complete but in more northerly forms the process is arrested after eight or occasionally only seven have been replaced on each wing. The inner first generation primaries thus last only a few weeks, whereas the outermost first generation feathers are retained until the full moult in the following year. In woodpeckers (Picidae) all of the primaries are replaced (the inner one or two may even be moulted before fledging) but the secondaries may be retained until the complete moult in the second year. In larger non-passerines, such as the Shag *Phalacrocorax aristotelis,* and eagles, the moult of the flight feathers in the second year may be incomplete and this results in highly complex patterns in subsequent years (see Part II).

THE SEQUENCE AND DURATION OF MOULT

THE SEQUENCE OF MOULT

The great majority of passerine species and many other species as well moult in a very similar sequence and this is for convenience referred to here as the 'basic sequence'. The two wings are usually in phase or very nearly so and the comments which follow therefore apply to both wings.

Primaries

Moult in the 'basic sequence' begins at the mid-wing with the dropping of the innermost primary (P1) followed by the adjacent primary (P2) and so on in a fairly regular order outwards towards the wing-tip — this is termed descendant moult. The new feather is normally partly grown before the next in sequence is dropped so that there are feathers of various lengths in the 'growth band' which progresses across the wing with old feathers ahead of it and new ones behind (Fig. 6). The number of feathers in growth at any one time depends on:

i. the rate of growth of the individual feathers.

ii. the length of the feathers.

iii. the length of interval between the start of growth of ajacent feathers.

Secondaries

In the 'basic sequence' moult begins with the outer-most secondary (S1) at the mid-wing and progresses inwards towards the body. This is termed ascendant moult. The innermost secondaries (or tertials), which number three in most passerines, moult separately from the other secondaries. Their moult sequence is variable but in species with three tertials the centre one frequently moults first followed by the inner and then the outer. The order is subject to considerable variation both within and between species (e.g. Ginn 1975).

Rectrices

The sequence of rectrix replacement is both very variable and often irregular. Perhaps the most frequent sequence is from the centre outwards in pairs (centrifugal), moult from the outside inwards towards the centre (centripetal) is also quite frequent and almost simultaneous moult of all rectrices often occurs in some species. Specialised and complex sequences are found, for example, in some tree climbing species in which the tail is used as a support. For a discussion of other sequences see Verheyen (1953).

Primary coverts

The primary coverts usually moult in exact sequence with their corresponding primaries, a notable exception being the Great Spotted Woodpecker *Dendrocopus major* — Part II.

Greater coverts

The secondary greater coverts do not moult in phase with their corresponding secondaries. The order of replacement is very variable both within and between species, but whatever the sequence, there are frequently many in growth at once, indicating short intervals between the start of moult in different feathers, and indeed it is not uncommon to find the entire tract at virtually the same stage of growth. In juveniles, if greater covert replacement is not complete it is the inner part of the row which is renewed with variable numbers of old feathers remaining at the outer end of the row.

Alula and carpal covert

Studies of the moult of the alula and carpal covert have been neglected until recently and most of the data collected by the B.T.O. has not been analysed. In general the moult of the carpal covert and alula follow on after the greater coverts.

Fig. 6

A. *Only the first two primaries have dropped.*

TERTIALS AND SECONDARIES									PRIMARIES										L/R
43	44	45	46	47	48	49	50	51	52	53	54	55	56	57	58	59	60	61	
0	0	0	0	0	0	0	0	0	3	1	0	0	0	0	0	0	0	0	R

B. *Five primaries, all the tertials and the first of the secondaries have dropped.*

TERTIALS AND SECONDARIES									PRIMARIES										L/R
43	44	45	46	47	48	49	50	51	52	53	54	55	56	57	58	59	60	61	
2	3	1	0	0	0	0	0	1	5	4	4	3	2	0	0	0	0	0	R

C. *There are only two old primaries and secondaries remaining: the new tertials are fully grown.*

TERTIALS AND SECONDARIES									PRIMARIES										L/R
43	44	45	46	47	48	49	50	51	52	53	54	55	56	57	58	59	60	61	
5	5	5	0	0	2	2	3	4	5	5	5	5	5	3	2	1	0	0	R

Three stages in the progression of moult through the wing of a hypothetical passerine.

RELATIONSHIP OF MOULT IN THE VARIOUS FLIGHT-FEATHER TRACTS

In passerines the duration of primary moult frequently spans the moult of all other feathers but recent work has shown that there are many exceptions (e.g. Flegg and Cox 1969). The secondary moult (S1-S6) normally begins at about the time that the sixth primary falls though this is subject to considerable variation both between and within species. On average secondary moult is completed at about the same time as that of the primaries but either group may finish first. The tertials generally moult shortly after the start of the primary moult. The timing of tail moult is much more variable but it usually starts during the early stages of primary moult and is completed before the primaries. As already noted, the primary coverts moult in phase with their corresponding primaries but the greater secondary coverts moult ahead of the secondaries. The alula usually moults during the later part of the primary moult. When greater primary and secondary coverts and alula feathers are in growth and particularly during the early "in-pin" stages, there is a danger of mistaking them for growing secondaries or primaries and this should be guarded against during moult recording by careful examination of the points of origin of each feather.

Exceptions to the basic sequence

(For details, see under each species in Part II).

Passerines

Exceptions to the basic pattern occur in the remex moult of the Dipper *Cinclus cinclus*, the primary, secondary and tail moult of the Spotted Flycatcher *Muscicapa striata*, and the tail moult of the Treecreeper *Certhia familiaris* (cf. woodpeckers).

Non-passerines: Primaries

In many non-passerine species the basic descendant primary moult sequence is followed but the intervals between shedding of feathers may vary, and the frequency of moult cycles may be such that one wave of moult has not reached P10 before a new wave starts at P1, so that there are several ages of feathers in the wing at the same time. Stresemann (1965) described this as '*staffelmauser*' or 'stepwise moult'. Sometimes the moult is irregular in that it does not progress sequentially along the wing, some feathers being passed by and renewed later, so that two growing feathers may be separated by one or more old or new ones — this is termed 'transilient' moult (Stresemann and Stresemann 1961a). Where there is a regular alternation of growing and non-growing feathers, as in the outer primaries of the Cuckoo *(Cuculus canorus)* this is termed "alternating" moult.

Various waterbirds have independently evolved virtually simultaneous moult of the flight feathers (primaries and secondaries) as already noted for the Mallard. In some other groups (e.g. *Falco* spp.) the primary moult begins in the middle of the tract at P4 or P5 and proceeds centrifugally, while the primaries of the Kingfisher *Alcedo atthis* are moulted in two groups concurrently.

Non-passerines: secondaries

The secondaries are often much more numerous than the primaries, can display far more variation and in many cases it may be very difficult to ascertain a sequence as such. There can be considerable variation between individuals and indeed between the two wings of the same individual. There are frequently several centres of moult from which replacement spreads. The 'tertials' sometimes moult as a unit (e.g. in waterfowl, Anseriformes), just as they do in passerines. The pattern of moult in the tail and the frequency or irregularity is much as in passerines.

FEATHER GROWTH

Heinroth (1906) described three phases of feather growth; firstly a slow initial phase confined to the feather germ in the base of the follicle; secondly a long phase of daily elongation during the major part of which the growth is more or less linear; thirdly a progressive slowing down of growth, a withdrawal of pulp from the calamus and the cornification of the feather base. Due to the problem of studying the first phase, most moult studies have necessarily been concerned with phases two and three, from the dropping of the old feather to the cornification of the base of the new (cornification being considered complete when there are no obvious signs of feather sheath remaining around the base of the feather). In the B.T.O. Moult Enquiry feathers are scored from 0 (old feather) to 5 (new feather fully grown), as illustrated in Fig. 7b.

Growth bars

As the flight feathers emerge from their sheaths, they often show distinct but faint 'growth bars' across their width (Michener and Michener 1938). It has long been considered that the sequence of these bars "indicate the growth of feathers and the regularity of feeding with its relative consumption" (Wood 1950). Often, in juvenile birds, some may be exaggerated causing 'fault bars' which can readily be seen. In extreme cases where the feather is so weakened at a bar that it may break it is likely that the bar results from a period of severe stress, such as a shortage of food. The more frequently found bars, which are formed by very slight differences in gloss or texture (but not colour) and are best seen in a glancing light, are probably the result of normal changes in metabolism and do not necessarily indicate a change in the rate of feather growth. Such metabolic changes occur in a diurnal pattern with night-time feather growth being based partly on the degradation and redistribution of body proteins while food intake during the day provides protein for feather synthesis as well as replenishing body protein stocks (Dolnik and Gavrilov 1979).

A knowledge of growth bars can be used as an aid the ageing some birds; as juvenile feathers are all grown together at the same time they may have corresponding growth bars throughout the plumage but certain groups of adult feathers may also be grown together (e.g. rectrices) and so the method needs to be used with caution (Svensson 1975). Green (n.d.) proposed that the rate of feather growth could be compared between tracts and individuals by measuring growth bars, but there has been little work on this topic.

The rate of feather growth and moult

As a very general rule, larger birds tend to take longer to complete their moult than smaller ones, partly because they usually have larger feathers but also because the feathers tend to be replaced in a less rapid sequence than in smaller birds (waterfowl, which moult their flight feathers simultaneously, are an obvious exception).

There have been relatively few studies which have measured feather growth rates and most relate to captive birds and thus may not be necessarily applicable to birds in the wild. Measurements of feather growth in free-ranging wild Barnacle Geese *Branta leucopis* showed that all birds had similar growth rates irrespective of age or sex (Owen and Ogilvie 1979). Work on finches has shown that although individual feather growth rates may vary between species, they are relatively constant within a species and thus differences in the overall rate of moult between individuals of the same species are associated with the number of feathers growing at one time (Evans 1966, Newton 1967, 1968). A study of waders in Morocco (Pienkowski *et al.* 1976) revealed that interspecific differences in the rate of moulting were related to differences in the rate of feather growth and that differences within one population of the same species were related to the number of feathers growing concurrently but that feather growth rates varied between different populations of the same species.

In a study of captive Bullfinches *Pyrrhula pyrrhula* Newton (1967) found that all of the primaries grew at a fairly constant rate of 3-4mm per day for much of the growth period but that the growth rate slowed down towards the end. The outer primaries slowed down their growth at an earlier stage than the inner primaries and thus took longer to complete growth than inner primaries of an equivalent length.

Seel (1976) studied the rate of deposition of feather material in five species of Corvidae and found that deposition rates varied during the course of a moult giving a sigmoid pattern, the reduced rates of deposition at the start and end of the moult being related to fewer feathers being in growth at these times.

RECORDING MOULT

The average field worker/ringer investigating moult requires a simple recording system to collect useful information. The recording system currently used in Britain and frequently elsewhere (e.g. Bancroft and Woolfenden 1982, Ford 1980, Noskov and Gaginskaya 1972), in which the remiges and rectrices are scored from 0 to 5 according to their age and stage of development (Fig. 7b) has been developed from the earlier work of Miller (1961), Ashmole (1962) and Newton (1966). As noted earlier (page 26) it has the disadvantage that no account is taken of the different length and mass of feathers and Seel (1976) has therefore criticised the technique for its imprecision. Seel studied the moult of five species of Corvidae by measuring the length of the individual feathers and relating these to their dry weights to produce a system of 'feather units', the dry weight of the whole plumage of each sex of each species being equivalent to 1000 units. The method, which is time consuming and requires the use of special equipment and dead birds, whilst being more precise than the traditional 'moult score', is beyond the scope and taste of most amateur ringers. However, further studies are clearly needed to investigate rates of feather growth and deposition of feather material.

An alternative method of studying primary moult is the collection of moulted feathers from nest or roost sites. This technique has been used, for example, in studies of accipiters (Opdam and Müskens 1976), Lapwing *Vanellus vanellus* (Snow and Snow 1976), Redshank *Tringa totanus* (Der Blanken *et al.* 1981), and various gulls and the Common Tern *Sterna hirundo* (Walters 1978, 1979). Information obtained by this method is valuable only if individual feathers can be safely identified and the same birds are being studied throughout the investigation (Appleton and Minton 1978).

Various methods have been used to describe the stages of and sequence of the moult of body feathers (e.g. Ainley *et al.* 1976, Ferns 1978, Norman 1981, Pitelka 1945, 1958), but at present there is no 'standard' method, and the scoring system used on the B.T.O. moult card is very imprecise. Seel's 'feather units' obviously relate to body feathers but, as noted above, are impractical for the amateur ringer. The lack of a suitable method for recording body moult is unfortunate since 'body' feathers account for more than half of the feather mass of a bird (see below) and their replacement may place proportionately more of an energy burden on the bird than the replacement of the remiges and rectrices.

Mean weights of feathers tracts in the Jay Garrulus glandarius *and the Bullfinch* Pyrrhula pyrrhula.

	Jay ♂ (10)*		Bullfinch (5)**	
Tract	wt	% total	wt	% total
	mg		mg	
Primary	8959	19.9	220	10.0
Secondary	3423	7.6	96	4.4
Tertial	1235	2.8	26	1.2
Rectrical	3210	7.2	136	6.2
Others	28037	62.5	1723	78.2

*after Seel (1976)
** after Newton (1966)

THE MOULT CARD

At a meeting in 1971 attended by ornithologists from 13 European countries it was agreed that a standard design of moult card should be used throughout the region. Being of a standard format and designed for easy transfer of data to computer files such a card would facilitate the exchange of data between countries for comparative studies (Ginn 1973).

The current B.T.O. card (Fig. 7a) is more comprehensive than earlier ones (Snow 1967), being applicable to both passerines and non-passerines, and includes space for recording additional parameters such as wing length and weight. Abbreviated instructions are given on the back of the card. Detailed instructions and explanations are provided separately with each batch of cards supplied.

Fig. 7a

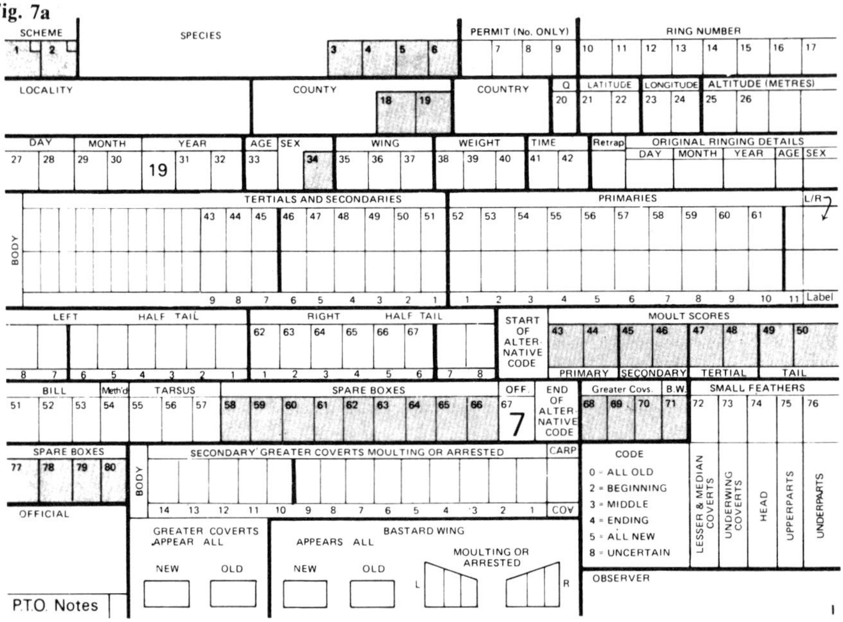

The moult card in current use in Britain. Small numbers in boxes refer to columns on the standard punched card, shaded boxes are to be coded in the office and not filled in by the observer in the field.

Fig. 7b

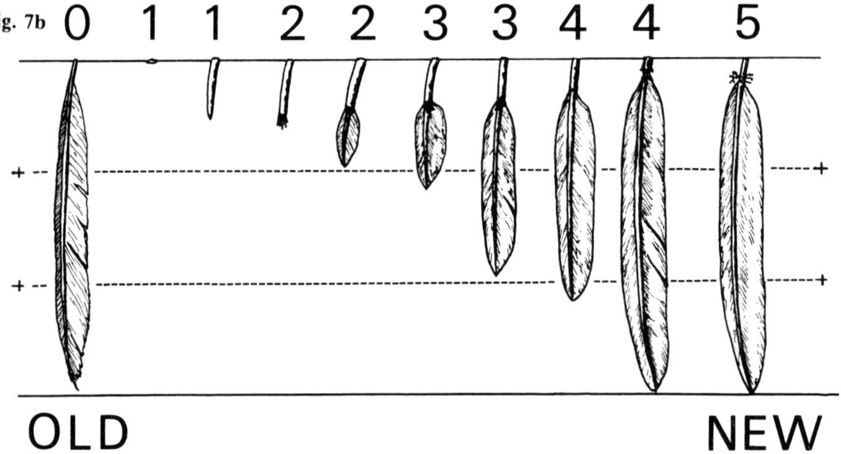

The feather scoring system. The dashed horizontal lines represent one third and two thirds growth. The numerical scoring system is:

0 *Old feather remaining.*
1 *Old feather missing or new feather completely in pin.*
2 *New feather* just *emerging from the sheath up to one third grown.*
3 *New feather between one and two thirds grown.*
4 *New feather more than two thirds grown and with remains of waxy sheath at its base.*
5 *New feather fully developed with no trace of waxy sheath remaining at base.*

Study of living birds

Before recording the details it is essential to account for the full number of primaries, thus at the same time establishing the division between the primaries and secondaries. In the early stages of wing moult there is usually no difficulty in finding the division, as the innermost primaries begin to moult sometime before the outer secondaries (Fig. 6a). But later, when both inner primaries and outer secondaries may be new and full grown (Fig. 6c), the division may not be obvious.

Since some primaries may be missing or difficult to see in early stages of growth, one should not simply count from the outermost feather and assume that when the right number of feathers has been counted one has reached the innermost primary; a check is needed. In many birds the primaries differ in shape or colour pattern from the secondaries; the latter are often squarer, or rounder and less pointed at the tip. This may be sufficient to establish the division. If the wing is gently opened and closed, the division will usually become clearer, as the primaries move as a unit and the outermost secondaries pivot over the innermost primaries. To confirm the correctness of the diagnosis, count both the primaries and the secondaries and see if both reach the expected numbers. One should not be satisfied until all the expected feathers are accounted for. But note that there is some individual variation, even in the number of primaries (though this is very rare — see page 11). Check any apparent abnormality with particular care and make a note on the back of the moult card stating that you are aware that what you have recorded is unusual, but that you have checked it and it is correct.

Study of dead birds and skins

For dead birds, proceed as for living birds; but in the case of non-passerines, especially species with large numbers of secondaries whose moult sequence is not well established, it may be worthwhile to pluck the coverts for better display of the flight feathers.

In museum skins the primaries are best examined from the ventral side and the secondaries from the dorsal side. The coverts can temporarily be pushed aside with a probe such as a pencil or knitting needle, and the downy base of the feathers, which obscure the flight-feather quills, can be laid with a drop of alcohol.

THE INTERPRETATION AND ANALYSIS OF MOULT DATA

The commonest questions asked about moult are likely to be:-

a) when is the moulting season.
b) what is the spread of the dates of start and finish of moult.
c) how quickly does a bird moult/what is the duration of moult in an individual bird.

The current B.T.O. recording system (page 28) lays emphasis on the remiges and rectrices as these are quickly and easily recorded with little variation between observers. Consequently most moult studies to date have related largely to moult in these tracts, especially the moult of the primaries since it has been widely regarded that, at least in most passerines, all of the other tracts are moulted within the time taken to replace the primaries (see Newton 1968). Subsequent studies have shown an increasing number of exceptions (e.g. Flegg and Cox 1969) but due to the ease of recording primary moult it remains the 'standard' measure of moult. Most moult studies thus seek to answer the above questions with respect to primary moult and not to moult in other tracts.

A scatter diagram of primary moult against date gives an easily understood pictorial representation of the progress of moult and is the usual first step in any analysis (for examples see Part II). From such diagrams one can determine the approximate timings of the start and end of moult and the spread in starting and finishing dates. With large samples of data it may be apparent that there is an uneven spread of records throughout the season. Passerine birds, which may drop up to four or five primaries over a very short period at the start of moult and thus suffer from reduced flight capability, frequently become more secretive at this time and are less likely to be caught in mist nets. A further factor which becomes apparent, especially when using data collected through the B.T.O. Moult Enquiry rather than from specialist studies by individual ringers or ringing groups, is that the start and especially the end of the moulting season are generally under-recorded.

The determination of the duration of primary moult is difficult. At the present time there is no wholly satisfactory method due to the fact that the data are usually insufficiently uniform to support the mathematical assumptions on which models are based. Retraps of ringed birds offer one way of determining the progress of moult in an individual bird but such records are generally insufficiently frequent to offer more than a guide line for interpretation of single capture moult scores by other methods. Studies of captive birds have the advantage that the birds can be examined often but the stress of being in captivity and of frequent handling may result in atypical moult (Newton 1967).

Early work on the Redpoll *Carduelis flammea* and the Bullfinch *Pyrrhula pyrrhula* indicated that the regression of primary moult score against date showed a straight line relationship (Evans 1966, Newton 1966). Newton (1966) expressed his surprise at this "since for any individual, a plot of primary score against date might be expected to give a curve of sigmoid shape, the primary score increasing more slowly at the start and end of moult when fewer than average feathers were growing. However, the composite scatter diagrams, obtained by plotting the primary scored of many individuals against date, show little trace of a sigmoid shape, probably because in the Bullfinch the period when fewer than average feathers are growing is short compared with spread in the date of onset of moult in the whole population". Following these initial studies it was widely regarded that similar straight-line relationships held more or less universally for all species, such lines being fitted to the data either by regression analysis or simply by drawing a line, judged by eye, through the scatter diagram.

Regression analysis seeks to predict one variable (the dependant variable) from another variable (the independant variable) on which the first variable usually depends. In the finch examples noted above, moult score was taken as the dependant variable and date as the independant variable, it seeming reasonable that moult score is dependant on date. This assumption is appropriate if one bird's score is considered but is inappropriate for data from a population.

Fig. 8

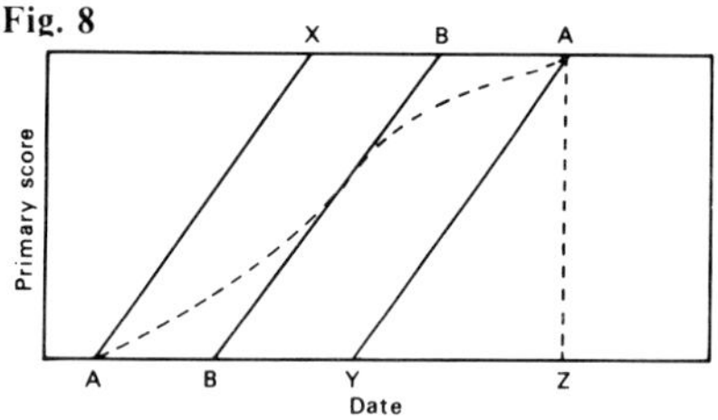

On this hypothetical scatter diagram, with all primary scores contained in the parrallelogram AXAY, the result of regressing date on moult score gives the line BB. A regression of moult score on date gives the sine curve represented by the dotted line between AA. Finish of moult for the whole population is given by Z.

Fig. 8 shows a plot of moult score against date for a hypothetical species, the parallelogram A-X-A-Y, enclosing the areas covered by the scatter diagram of moult scores. In Fig. 8 a regression of moult score (dependant) against date (indepentant) gives the average stage of moult of the individuals at the time of capture (line A-A), this being a function of both the duration of moult of the individual bird and the time period over which the population starts to moult. This line is sigmoid in shape rather than straight due to the fact that the distribution of moult scores for a particular date is not 'normal' since it is not possible to have a moult score of less than 0 or more than 50 (45 in the case of species with only 9 primaries). Furthermore in species such as many of the waders where the timing of arrival before, and departure after moult is protracted within the population, this results in mean moult scores for the population being 'depressed' by late starting birds. If date (dependant) is regressed against moult score (independant) the resulting line (B-B) approximates much more closely to the progress of moult of an individual bird, the variance about this line being a measure of the variability in the starting and finishing dates. For this method to be truly representative it is necessary to have an even spread of records throughout the season (and hence a 'normal' distribution of dates for a particular moult score). Such uniformity can seldom, if ever, be achieved with wild birds since, as noted above, many species are difficult to catch during the early stages of moult. However if a numerical analysis is required, regression of date on moult score usually results in a line similar to that drawn 'by eye' and with a slope similar to that shown by retraps.

It is clear from Fig. 8 that the two methods of regression may give greatly different results and it is most important to bear this in mind when referring to the literature and to remember not to compare directly, moult duration figures for different populations obtained by different methods (Pimm 1976).

A variety of other methods have been used in attempts to overcome the problems outlined above, including the use of median moult scores (Appleton and Minton 1978) and probit analysis (Johnson and Minton 1980).

A further complication is that in birds with long outer primaries an increase of one unit in moult score may represent a much greater amount of new material in a long feather than in a short feather; this is particularly marked for example in many shorebirds where the long outer primary may weigh more than three times as much as the small inner primary. Thus if one assumes that the rate of increase in moult score is uniform (i.e. a straight line) with respect to time, this implies that the rate at which feather material is laid down increases towards the end of the moult period — an assumption which does not apply to all species (Newton 1966). This problem can be reduced by converting moult scores to feather mass (Summers *et. al.* in prep.). The detailed methodology of Seel (1976) which relates feather growth to feather 'units' overcomes this problem but is impractical (see page 27).

At the present time mathematicians are still actively considering methods of analysis for moult data, and detailed analysis remains the realm of the mathematical ecologist rather than that of the amateur bird ringer. For general purposes, scatter diagrams and lines fitted by eye are the easiest and are generally satisfactory.

PART II : THE SPECIES ACCOUNTS

INTRODUCTION

The following section summarises the information currently available concerning the moult of British birds, both from completed cards in the B.T.O. collection and from published sources. Each account is intended to be largely self-contained so that all available information for a particular species is in one place. However a brief introductory note to certain orders and families is included at the head of the relevant sections.

Where possible the species accounts are based on British material. For species where local material is lacking or incomplete, material from elsewhere has been used, and this has also been included in some species for comparative purposes.

Due to the limited space available the species accounts are necessarily very condensed, and in many cases give a simplified picture of what are frequently complex moult patterns. Key numbers are given to references in the bibliography to enable the reader to refer back to the original sources of information. The bibliography, which includes some studies on 'British' species conducted outside Britain, is extensive but does not claim to be exhaustive. It should be noted that much of the work conducted in East Africa probably relates to East European populations and may not, be directly applicable to British populations.

A species has been included if it has bred in Britain in recent years or, if breeding elsewhere, regularly undergoes a major moult (including some/all remiges and rectrices) in Britain. This has reduced the number of species covered to 239. It will undoubtedly be argued that some 'border-line' cases could have been omitted, whilst others perhaps deserve to be included, but this is an inevitable consequence of limited space. It should be noted that information on moult of all European passerines has recently been summarised by Kasparek (1981), and information on moult for all species is summarised by Cramp and Simmons (1977, 1980, *et seq.*) and Bauer and Glutz von Blotzheim (1966 *et seq.*). The work of Stresemann and Stresemann (1966) contains much valuable information, but their interpretations of moult patterns must be treated with some caution since these are often based on museum material from widely scattered localities.

Explanation

i. The English names are those in common use in Britain. Scientific names and the sequence of families and species are those of Voous (1977). The asterisks beside the scientific names indicate the number of moult cards in the BTO collection:

*	0-10 cards
**	11-100 cards
***	101-1000 cards
****	1000 + cards

ii. The number of primaries, secondaries, tertials and tail feathers (rectrices) are taken from Snow (1967), ammended where necessary in the light of recent work. Very occasionally birds may be found with an 'extra' primary or with one 'missing' (page 11) and such cases should be checked very carefully. Irregularities in the number of tail feathers are more frequent. Snow (1967) noted the general lack of studies concerning the number of secondaries in birds, the problem of determining where the secondaries finish at the inside of the wing, and the fact that there may be variation in the number of secondaries within a species. These factors have resulted in published figures for a species not always being in agreement, and in such cases a range of figures is given. For species where more than one figure is given the less frequently encountered number is placed in parent-heses ( ). At the present time there still remain many species for which the number of secondaries has not been determined and these are indicated by a dash "-" in the accompanying texts. The letters "E" and "D" refer to whether the species is eutaxic or diastataxic — in the latter group the fifth

Comparisons of different plumage and moult terminologies

Cramp and Simmons (1977)	Dwight (1900)	Humphrey and Parks (1959)
Down	Down	Down
juvenile moult	*postnatal molt*	*postnatal molt*
Juvenile	Juvenal	Juvenal
postjuvenile moult **(PJ)***	*postjuvenal molt*	*prebasic I molt*
1st immature non-breeding	1st winter	Basic I
1st pre-breeding moult	*1st prenuptial molt*	*prealternate I molt*
1st breeding	1st nuptial	Alternate I
post-breeding moult **(PostN)**	*1st postnuptial molt*	*prebasic II molt*
Non-breeding (NNP)	2nd (adult) winter	Basic II
pre-breeding moult **(PreN)**	*2nd prenuptial molt*	*prealternate II molt*
Breeding (NP)	2nd (adult) nuptial	Alternate II

* The abbreviations in parentheses are those used in the species accounts. Moults between plumages are shown in italic. A more complex terminology is needed to describe the moults of large species with long pre-breeding periods whose plumage may not reach definitive adult status for many years.

secondary appears to be 'missing' (page 101). Other abbreviations used are:

exc.	exceptionally	o.r.	outermost reduced
fre.	frequently	r.	reduced
m.	minute	v.	vestigial
m.r.	much reduced	v.r.	very reduced
o.m.r.	outermost much reduced		

iii. The terminology of plumages and moults is generally based on Cramp and Simmons (1977), who follow Dwight (1900) and Amadon (1966) — see also Stresemann (1963). However to reduce confusion in the abbreviations the term 'nuptial/nesting' (N) has been used in preference to 'breeding' (B). The classification proposed by Humphrey and Parkes (1959, 1963) is followed by many American authors including Palmer (1969 *et seq.*) and the terminologies are compared in the table to assist readers wishing to consult the American literature.

iv. In all species primary moult is descendant and secondary moult ascendant unless stated otherwise.

v. The moult seasons given in the text cover the main moult periods when the majority of the population can be expected to be undergoing moult, months given in parentheses ( ) indicate that a few birds are found moulting at these times. It is important to remember that many factors may affect the timing of moult, both in an individual and in the population as a whole. In Britain, where most species moult after breeding, annual variations in the timing of breeding may result in similar variations in the timing of moult. Diseased, starving and injured birds may moult later than healthy birds (Wilson 1910) or in irregular sequences (Ginn and Glue 1974, Hardy *et al.* 1981). Heavy infestations of Malophaga may also result in unseasonal moult (Taylor 1981).

vi. Estimates given in the text for the duration of primary moult are taken largely from the literature and may be based on recaptured birds, lines fitted by eye to scatter diagrams, or on regression analyses (moult score on date (incorrect), or date on moult score) and are thus not directly comparable and should be taken only as a rough guide. For a discussion of methods for analysis of moult scores see page 29 .

vii. The scatter diagrams of primary moult score against date are drawn from data in the B.T.O. collections unless otherwise stated. Sample sizes are indicated in the top left of the figures. The time (x) axis is marked in seven-day periods and months are indicated. Since different species moult at different times of the year the start of the time scale is not uniform in all figures. Primary moult score is marked on the 'y' axis, with a maximum score of 45 or 50 depending on whether nine or ten primaries have been scored for these species. For those species with substantially more than 100 records envelopes have been drawn round the points, the enclosed area being shaded and outlying points shown individually. For species with fewer data points are shown individually; the solid lines joining points indicating the progress in moult score of bird caught more than once during the same moulting season. In all figures lines have been fitted 'by eye' to suggest a possible 'average' moult duration (see page 30), and the number of days as estimated by the lines are indicated at the top of them. The shaded figures (unless otherwise indicated) include data up to and including 1977, the remaining figures include data up to and including 1981.

viii. N.B. In some passerine species adults finish secondary moult after the primary moult is finished. At these times there is a risk of confusion with birds undergoing post-juvenile moult.

Abbreviations and terms used in the Species Accounts

1W	1st winter	
1S	1st summer	
2S	2nd summer	
1Y	1st year — Euring code 3	These years are calendar years with age change overnight 31 Dec./1 Jan.
2Y	2nd year — Euring code 5	

A Alula (bastard wing)
Ad Adult
Ascendant refers to direction of moult — in this case towards the body
Av Average
B Body
BP Brood patch
C Circa (about)
Complete refers to the extent of the moult — includes remiges and rectrices
Descendant refers to direction of moult — in this case away from the body
Est Estimated
Finish refers to the ending of the moult
FNP Full Nuptial (breeding) Plumage
GC Greater wing coverts
Imm Immature
Juv Juvenile
LWC Lesser Wing Coverts
NNP Non-breeding (-Nuptial) Plumage
NP Nuptial (breeding) Plumage
P Primaries (Primary moult)
P1, P2 Primary 1, primary 2 etc.:— numbered descendantly (from the body outwards)
Partial refers to extent of moult — not complete and generally not involving primaries or many of the secondaries
PJ Post juvenile (moult)
PostN Post-nuptial (after breeding) moult
PreN Pre-nuptial (before breeding) moult
R Rectrices — main tail feathers (Rectrix moult)
R1,R2 Rectrix 1, rectrix 2 etc.:— numbered from centre of tail outwards in pairs (R6 on bird with 12 tail feathers refers refers to the outermost pair)
S Secondaries (Secondary moult)
S1,S2 Secondary 1, secondary 2 etc.:— numbered ascendantly (towards the body)
SubAd Sub-Adult plumage
T Tertials (inner secondaries — in this work generally used only with reference to passerines)
T7,T8 Tertials are referred to by their number in the Secondary sequence: for most passerines T7 is the outermost and T9 the innermost.
WC Wing Coverts

NON-PASSERINES

GAVIIDAE

RED-THROATED DIVER **Gavia stellata*

P:10+1 m. S:23,D R:18

Ad PostN complete; flight feathers simultaneous, starting (late Aug) late Sept-Dec. **Ad PreN** partial (B,R, some LWC) starting Feb-early April; first in NP mid April. **PJ** partial (B,R, some WC), some start Dec. others not until Feb. Imm Spring partial (much B, some R) May-late July, starts before end of PJ. Imm Autumn complete, P and S simultaneous in summer. **(111,430, 581).**

BLACK-THROATED DIVER **G. arctica*

P:10+1 m. S:-,D R:16-18

Ad postN partial (B,R, some WC); some start Sept. others Dec. **Ad preN** complete from Jan., P and S simultaneous Feb-April; one unaged Isle of Man late March all P stage 1. **PJ** gradual replacement of B Jan.-summer; P and S simultaneous July-Aug. In 3rd year preN complete Feb.-May later than Ad, P and S April-May. **(111,348,350, 523,524,714).**

GREAT NORTHERN DIVER **G. immer*

P:10+1 m. S:-,D R:20

Ad postN partial (B,R, some WC); begins late July-early Oct., finished Oct.-Jan. **Ad preN** complete starting Feb.-March, finished April-May; P simultaneous Feb.-April; 3 UK records in early March with all P at stage 0,2,3. **PJ** birds in 2nd year almost continuously moulting; R and part B from Feb.-late spring; P and S simultaneous in summer. Partial moult in autumn, complete in spring of 3rd year, probably later than Ad preN. **(111,348,581,714).** WHITE-BILLED DIVER *Gavia adamsii* moult apparently similar to Great Northern Diver. **(430).**

PODICIPEDIDAE

LITTLE GREBE **Tachybaptus ruficollis*

P:11+1 m. S:-,D R:v.

Ad postN complete, timing variable depending on end of breeding; P and S simultaneous July-Oct. (most Aug.-Sept.), flightless 3-4 weeks; B gradual Aug.-Nov. (rarely Dec.). **Ad preN** partial (B,R, some inner S, some inner WC) Jan.-April (variable, later in cold springs). **PJ** partial (B,R) not sharply separated from juvenile moult, timing variable, early birds finish Sept., some still moulting Dec. **(111,581, 650).**

GREAT CRESTED GREBE **Podiceps cristatus*

P:11+1 m. S:21-22,D R:v.

Ad postN complete July-Dec. P and S simultaneous Aug.-Oct., flightless 3-4 weeks; B starts when P c$\frac{1}{3}$ grown; ♂ usually 2-3 weeks before ♀. **Ad preN** partial Dec.-late spring, traces of tippets appear first; not sharply separated from postN and moulting almost continuous, most B (except white underparts) renewed twice a year. **PJ** partial (B but not white underparts or longer scapulars, some WC) Jan.-late spring. **(111,313,365,496,539,581,663).**

RED-NECKED GREBE **P. grisegena*

P:11+1 m. S:20,D R:v.

Ad postN complete, P simultaneous July-Sept., but one Sussex 15 Nov. all P and S stage 1. **Ad PreN** partial (B,R, inner S, WC?) Dec.-May. Birds still in NNP in Feb. may be 2Y. **PJ** partial (B) Sept.-late Jan., timing variable, some nearly finished Nov., others just starting. **(111).**

SLAVONIAN GREBE **P. auritus*

P:11+1 m. S:-,D R:v.

Ad postN complete; P simultaneous, ♂ Aug.-Sept., ♀ probably Oct.; B starts early June-mid July, ♀ averages c1 month earlier than ♂. **Ad preN** partial (B, some LWC) March-April. **PJ** partial (much of B) starting (Sept.) mid Oct.-end Nov., finishing late Jan. 1st PreN starts soon after finish of PJ, early March-mid May. **(111,169,581).**

BLACK-NECKED GREBE **P. nigricollis*

P:11+1 m. S:18-19,D R:v.

Ad postN complete from mid June, Ads with young July-mid Sept., P simultaneous, Sept.-Nov.; many in NNP Sept., others not before Nov. **Ad preN** partial (B,R) Feb.-April, some birds (2Y ?) not in NP until

May. **PJ** partial (B,R) summer-Dec., apparently starting soon after end of juvenile moult. **(111,186,581).**

PROCELLARIIDAE

FULMAR **Fulmarus glacialis*

P:10+1 m. S:20,D R:14

Ad postN complete; timing of start of P moult uncertain; formerly thought to start after departure from breeding colony, late Aug. after successful nesting, earlier if failed; recent Scottish observations show Ad start when with chick, non-breeders and failed breeders earlier (mid-June); on St Kilda moult c4 weeks earlier than at other British colonies (R. Furness pers. comm.). P lost and replaced in rapid succession occasionally with up to 8 growing simultaneously (some may even become flightless), but in final stages P grow slowly not reaching full length before late Feb., B starts before wing (July) and continues until mid Feb. with feathers of 2 successive generations present for most of year; R renewed simultaneously with outer P . **PJ** complete May-Sept. of 2nd year, probably becoming progressively later in subsequent years. **(44,94,111,216,363,581,715).**

CORY'S SHEARWATER *Calonectris diomedea*

P:10+1 m. S:-,D R:12

Ad postN complete, timing unknown but some in moult Sept. when most not started; B July-March, starting when eggs hatch; moult slow. **(111,287).**

GREAT SHEARWATER *Puffinus gravis*

P:10+1 m. S:c21,D R:12

Ad postN complete, end May-mid Aug. in winter quarters; P moulted in rapid sequence with up to 6 growing simultaneously but never flightless; B moulted at same time as P, but R somewhat later; S moulted in 4 groups: S1-S4, S5-S12/S13/S14, S19/S20-S21/S22 ascendantly, and S18/S19-S14/S13/S12 descendantly; whole moult lasts c90 days, growing P increase in length c5mm/day. **PJ** partial (B) July-Aug. 1st year. In 2nd year complete moult earlier than Ads, starting late March-April when still in South Atlantic. Birds in North Atlantic in wing moult when Ads breeding probably older Imms. **(111,340, 353,363,402,503,585,666,715).**

SOOTY SHEARWATER *P. griseus*

P:10+1 m. S:20,D R:12

Ad postN complete May-Aug.; Imms presumably earlier; records of birds in moult from Feb., some still in moult in Nov. Pattern of moult similar to Great Shearwater. **(111).**

MANX SHEARWATER *P. puffinus*

P:10+1 m. S:c22,D R:12

Ad postN complete, Atlantic *puffinus* main moulting area in South American winter quarters; B starts July-Aug, P Sept., finish Feb.-April. Non-breeding brids may moult earlier. Balearic *mauretanicus* moults June-Oct. (P, S, B start at same time, R later). S moulted in 4 groups: S1-S4, S5-S12/S13/S14, S19/S20-S21/S22 ascendantly, and S18/S19-S14/S13/S12 descendantly. R irregular. **(111,349).**

HYDROBATIDAE

STORM PETREL *Hydrobates pelagicus*

P:10+1 m. S:-,D R:12

Ad postN complete; wing and R start during 2nd ½ incubation, c70% have P moult on peak fledging date; B starts during incubation; moult of B and wing continues slowly in winter quarters, some still growing outer P on arrival at breeding grounds and c½ still moulting S. Pattern of S replacement complicated, often suspends. **PJ** complete, B starts midwinter, P start c.June, finished Oct-Nov. **(111,353,525).**

LEACH'S PETREL *Oceanodroma leucorhoa*

P:10+1 m. S:14,D R:12

Ad postN complete, B starts at end of breeding, R and P soon after, mainly Oct.; main moult in winter quarters Nov.-Feb.; some nonbreeders start P Aug.-Sept?; 20 birds in East Atlantic (off Portugal/West Africa) between 11 Oct. and 24 Nov. had av. Pscore 35, range 14 (24 Nov.) — 49 (22 Nov.); one Somerset 26 Oct. Pscore 16, one inland Kent 14 Dec. Pscore 22. **PJ** complete, P start cApril of 2nd year, finished Oct.-Dec. In California, USA P start c70 days after B, S start at least 30 days after P, R start c40 days after B and finished in c75 days. **(4,111,353,581).**

SULIDAE

GANNET **Sula bassana*

P:10+1 m. S:28(26-31),D R:12

Ad postN 'complete'; P serially descendant with 3 active centres in wing, July-Dec. (occasionally Feb.); P usually suspended from before egg laying until after hatching, a few late breeders resume P moult whilst still incubating; S and R may be dropped from mid April in breeding birds; R irregular. **PJ:** gradual B moult from Jan. 2nd year (occasionally from Nov. 1st); details of P moult unknown but 1st series reaches P6/P7 in Sept. of 2nd year and 2nd series starts with P1 (see Shag). **(111,399,400).**

PHALACROCORACIDAE

CORMORANT **Phalacrocorax carbo*

P:10+1 m. S:-,D R:14

Ad postN 'complete'; P serially descendant, each month July-Dec. c1 P of each moulting series replaced; 2 (rarely 3) active series in each wing; R irregular, more or less alternating, mainly June-Dec., B apparently completely moulted at same time. **Ad preN** partial (head, neck, some B) Jan.-April. **PJ** partial (head, neck, part B), Aug.-Dec., much variation and effectively from Aug. 1st year — Dec. 2nd year B moult almost continuous; Complete moult 2nd year, B Feb.-Dec., wing and R June-Dec.; P slow, by Dec. P1-P6 replaced, some R often retained. Spring of 3rd year 1st P series resumed and 2nd series starts at P1, Juv P10 may be retained until spring of 4th year. **(111,581).**

SHAG ***P. aristotelis*

P:10+1 m. S:18-19,D R:12

Ad postN 'complete'; P serially descendant starting June-Aug., 1 P replaced per month, suspended Nov., resumed from Feb. at slower rate, speeding up after breeding when new series starts at P1; usually 2 active series in each wing, occasionally 3 when P10 replaced after 2 winters; head and neck

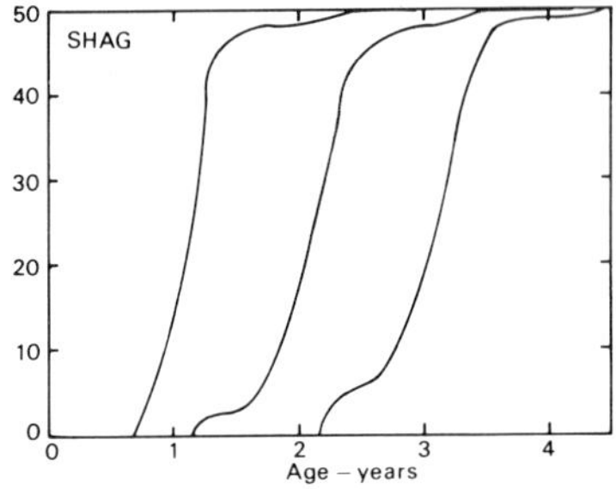

Progress of the first three complete moults of primaries in a Shag: left – **PJ,** *middle – first* **postN,** *right – second* **postN.** *Redrawn from Potts (1971).*

moult from June, rest of B from Aug., finished Jan. **Ad preN** partial (head, neck, part B) overlaps with postB, Nov.-Feb. **PJ** 'complete' lasting over 1 year; P serially descendant from May of 2nd year (sometimes earlier, Feb.), many finished by Dec. but 33% ♂ and 70% ♀ suspend and retain P10 until June of 3rd year, and 2% ♂ and 4% ♀ retain P10 until summer of 4th year; B starts Sept., suspended during winter, resumes next spring, most finish late autumn of 2nd year; R irregular throughout 2nd year; WC moult with B, GC with corresponding remiges. PostImm complete, wing from summer of 3rd year. **(44,111, 479).**

ARDEIDAE

BITTERN **Botaurus stellaris*

P:10+1 m. S:-,D R:10

Ad postN complete July-Jan., most finish Nov. a few not until March; one in UK replacing P1 and P2 22 Aug. Sequence of P replacement apparently irregular but poorly known. **PJ** partial (B,R) July-Nov., head apparently moulted again mid-winter? **(111,581).**

GREY HERON **Ardea cinerea*

P:10+1 m. S:19,D R:12

Ad postN complete June-Nov., begins before young fledge; inner P descendant, outer P irregular, P10 often moulted before P8 or P9; in UK 6 Sept. Pscore 24, 27 Sept. Pscore 19, 3 birds in apparently suspended moult: 23 Feb. replaced P1-P3, 3 March replaced P1-P6, 18 April old P9 retained, also old S4 and S7-S10; B may continue into winter; R centripetal. **PJ** partial (most of B, some WC) Sept.-Feb., timing highly variable. Imm moult similar to Ad postB but P may start earlier. **(111,374).**

ANSERIFORMES

Moult patterns are frequently complex and the extent to which moults are undergone in any particular feather tract may vary considerably within a species. A number of species (e.g. Shoveler *Anas clypeata)* appear to have a 'supplemental' plumage between 1st Non-Breeding Plumage and 1st Breeding Plumage. Most of the information in the species accounts is taken from Cramp and Simmons (1977) where the reader will find additional details. General discussions on moult patterns are given by Palmer (1972) and Weller (1976). Periods of flightlessness should be regarded with some caution since birds can fly before the primaries are fully grown and it is not always clear from the literature whether periods quoted refer to actual flightlessness or to time required to grow full length Primaries (e.g. see Mute Swan *Cygnus olor*). Many species undergo a 'moult migration' to sites which offer food and safety from predation during the flightless period (see e.g. Kumari 1980, Owen and Ogilvie 1979, Salomonsen 1968).

ANATIDAE

MUTE SWAN **Cygnus olor*

P:10+1 r. S:22-28,D R:20-24

Ad postN complete, flight feathers simultaneous; flightless c6-8 weeks **(111),** in USA 3-4 weeks **(691)**, the apparent inconsistency probably explained by Danish study where birds could fly 35-42 days after start of P moult but P not fully grown for 66-67 days in ♂, 60 days in ♀ **(345)**; ♀ moults when young small, ♂ when ♀ almost able to fly again; B and R moulted when wing replaced. **PJ** partial (B) from shortly after fledging-autumn; complete moult from June of 2nd year. Non-breeding birds moult mid June-Aug.; ♂ before ♀. **(7,71,111,299,345,375,546,691).**

WHOOPER SWAN **C. cygnus*

P:10+1 r. S:22-28,D R:20

Ad postN complete, flight feathers simultaneous, flightless c5-6 weeks, June-late Aug.; ♀ moults wing before ♂; B starts when wing in quill; R moult prolonged finished Dec. **PJ** partial (B) Oct.-March, may include a few T, innerWC and R. Other Juv. feathers replaced completely in summer of 2nd year (occasionally retain some Juv. WC), sequence as Ad postB. **(111,299).**

GREYLAG GOOSE **Anser anser*

P:10+1 r. S:-,D R:18(rarely 20)

Ad postN complete, flight feathers simultaneous, flightless c1 month mid May-mid Aug.; non-breeders earlier than breeders; WC moulted with remiges, rest of B and R moulted when remiges replaced; R moult prolonged, alternate. **PJ** partial (B, some centre R), variable, some have mostly new B (but not all R, or any WC) by Oct., others not until Jan. or later. **(111,147).**

CANADA GOOSE **Branta canadensis*

P:10+1 r. S:17-18,D R:16-18

Ad postN complete, flight feathers simultaneous, flightless c3-4 weeks; B starts soon after young hatch, finished after wing; Yorkshire and Beauly Firth moulting flocks flightless mid/late June-end July. **Ad preN** partial (head, neck) spring. **PJ** partial (head, B, R) from early autumn. **(111,668).**

EGYPTIAN GOOSE **Alopochen aegyptiacus*

P:10+1 r. S:-,D R:14

Moults little studied. In South Africa wing moult related to breeding (and local rainy seasons); flightless c30 days, can fly when P c70% total length; B and R start shortly after breeding. **PJ** partial (B,R,T, variable number of WC), starts soon after fledging and finished within a few months; wing moulted at same time as Ads. **(111,531,534, 535).**

SHELDUCK **Tadorna tadorna*

P:10+1 m. S:20,D R:14

Ad postN complete, flight feathers simultaneous, flightless 25-31 days, early July-mid Oct.; B starts shortly after hatching of young, R at same time as wing. Most British birds moult on the West German Grosser Knechtsand, Elbe-Wesser Estuary, but smaller moulting flocks at Bridgewater Bay, Somerset, Firth of Forth, the Humber estuary and the Wash. Ads left to tend crêches moult locally. **Ad preN** partial (B) Aug.-Dec. **PJ** partial (B) from fledging, late July-early Oct.; from late Sept. moult to 1st ImmNP, timing variable, moult slows or suspends during winter. 1st postN like Ad postN. **(89,90,111,154,197, 262,420,465,509,541,614).**

MANDARIN **Aix galericulata*

P:10+1 r. S:-,D R:16

Ad postN complete, flight feathers simultaneous; ♂ T first, then B mid May-July, R July; flightless c1 month May-early Aug.; ♀ probably as ♂ but wing starts c1 month later, one ♀ Surrey 1 Sept. Pscore 47 (P9 and P10 still growing). **Ad preN** partial (B,T), ♂ start late Aug., finish Sept.-Oct., ♀ later. **PJ** partial (B,R), attains 1st ImmNNP from Aug., often finished late Sept., when first feathers of 1st ImmNP appear; 1st preN partial (B,T). **(111).**

WOOD DUCK **A. sponsa*

P:10+1 r. S:-,D R:16?

Ad postN complete, flight feathers simultaneous, flightless 22-24 days, early July in ♂, early Aug. in ♀ (USA); ♀ may start B, and one or more S, during incubation as early as mid May; ♂ starts mid May and attains NNP by early July. **Ad preN** partial (B,T), ♂ begins July-early Aug., ♀ c1 month later. **PJ** partial (B) late July-early Oct. (starts before flight feathers fully grown), partial moult to 1st NP starts when c100 days old. **(430,431).**

WIGEON **Anas penelope*

P:10+1 r. S:16 (S12-S16=T),D R:14-16

Ad postN complete, flight feathers simultaneous; wing late June-early Sept. in ♂ and unsuccessful ♀, successful ♀ late July-Sept.; B July-early Sept. **Ad PreN** partial (B,T, centre R) starts when wing finished, ♂ earlier than ♀, most in NP by Dec. **PJ** partial Sept.-Oct., 1st preN m6ult partial (B,R,T) from Oct. in ♂, ♀ later usually from Dec. some earlier, **(111,273,515,541,570).**

GADWALL **A. strepera*

P:10+1 r. S:-,D R:16 (15-19)

Ad postN complete, flight feathers simultaneous, flightless c4 weeks, ♂ June-late Aug, ♀ later, sometimes not until Oct.; ♂ head and B May-July, ♀ June-Aug. (earlier in USA). **Ad preN** partial (B,T, most/all R), ♂ in almost FNP by Sept., ♀ by Oct. **PJ** partial Aug.-Oct.; 1st preN from Sept. (B,T,R), most finish by Dec. (by Jan. in USA). **(111,273,425,515,541,570).**

TEAL **A crecca*

P:10+1 r. S:15 (S12-S15=T),D R:16

Ad postN complete, flight feathers simultaneous, flightless c4 weeks, ♂ July-late Aug., ♀ mid July-late Sept.; head and B ♂ early June-late July, ♀ later. **Ad preN** partial (B,T, variable number R), ♂ Sept. Nov., ♀ from late Sept., sometimes not finished until Feb.-March. **PJ** partial, Aug.-Sept. in ♂, Aug.-Oct. in ♀. 1st preN partial B,T,R) Sept.-spring, ♂ largely finished by Nov, ♀ usually later seldom renewing T until spring. **(111,273,515,541).**

MALLARD ***A. platyrhynchos*

P:10+1 r. R:18-20

S:17/18 (S12-S17/18=T), D

Ad postN complete, flight feathers simultaneous, flightless 28-37 (captive birds flightless c29-33 days), ♂ early June-late Aug. (most mid July), ♀ start when young independant late June to late Sept. Able to fly when P 78-79% full length, P grow at c4.5mm/day. ♂ start B late May, successful ♀ when young c2 weeks old. **Ad preN** partial (B,T,R), starts shortly after end of wing moult, ♂ Aug.-early Oct., ♀ usually finish before Nov. **PJ** partial (B) July-Sept, may start before P fully grown. 1st preN partial (B,T,R), ♂ Aug.-Dec., ♀ slower, both sexes may not finish until spring. **(46,111,266,273, 426, 515,570,716).**

PINTAIL **A. acuta*

P:10+1 r. S:-,D R:16

Ad postN complete, flight feathers simultaneous, flightless c4 weeks, ♂ July-mid Sept., ♀ early Aug.-mid Oct. **Ad preN** partial (B,T,R), ♂ in complete NP by Nov., ♀ by Dec. **PJ** partial (B). 1st preN partial (B,T,R), ♂ in NP by Dec, ♀ shows some NP Oct.-Jan., most attained in spring. **(111,273,515,541,570).**

GARGANEY *A. querquedula*

P:10+1 r. S:-,D R:14

Ad postN complete, flight feathers simultaneous, flightless 3-4 weeks, ♂ mid June-mid Aug., ♀ up to c1 month later. B late May-July in ♂, ♀ later after young independent, finish as late as end Aug. **Ad preN** partial (B,T,R), starts slowly after renewal of flight feathers, finished in winter quarters Nov.-early March. **PJ** partial (B) from Aug., or shortly after arrival in winter quarters. 1st preN partial (B,T,R) from Dec. **(111,273,515).**

SHOVELER *A. clypeata*

P:10+1 r. S:-,D R:14

Ad postN complete, flight feathers simultaneous, flightless 3-4 weeks, ♂ mid June-mid Aug., ♀ late July-early Sept.; ♂ B early May-early June, ♀ c1 month later. **Ad preN** partial (B,T,R) Aug.-Dec. in ♂, ♀ Sept.-Feb./March. **PJ** partial (B) Aug.-Sept. in ♂, Sept.-Oct. in ♀. 1st preN partial (B,T,R), ♂ Sept.-Dec., ♀ finish Feb.-April. **(111,273,515,541).**

POCHARD *Aythya ferina*

P:10+1 r. S:-,D R:14

Ad postN complete, flight feathers simultaneous, flightless c3-4 weeks, ♂ late June-early Sept., (mainly late July), ♀ early July-late Oct. (mainly late Aug.); ♀ moults most of B late Feb./March-June, ♂ June-July. **Ad preN** partial (B) Sept.-Oct., ♂ in NP from late Oct., ♀ c1 month later. **PJ** partial ♂ from July, 1st preN partial from Sept., variable; ♀ moult less complete. 1st postN like Ad, ♀ in spring, ♂ in summer, flight feathers retained until summer. **(111,273,422,642).**

TUFTED DUCK *A. fuligula*

P:10+1 r. S:-,D R:14

Ad postN complete, flight feathers simultaneous, flightless 3-4 weeks, ♂ late June-early Sept., ♀ 1-2 months later; ♂ B late May-early July, ♀ March-May; ♀ may replace R once or partly twice April-Sept. **Ad preN** partial, ♂ starts when P full grown, usually finishing Nov., ♀ starts Aug., finishes Oct.-early Feb. **PJ** and 1st preN partial (B,T,R), highly variable; PJ from Aug, 1st preN from Nov.-April in ♂; some ♀ show little NP and moult to 2nd NNP in spring. **(111,273,422,541,570).**

SCAUP *A. marila*

P:10+1 r. S:-,D R:14

Ad postN complete, flight feathers simultaneous, flightless probably 3-4 weeks. In Denmark ♂ and ♀ moult at same time Sept./Oct., the moult migration being part of the autumn migration as 1Y birds accompany moulting ♀. In USA **Ad preN,** ♂ intense B moult Nov/early Dec, some in moult Jan./Feb./March, uncertain whether this is a continuation of earlier moult or resumption after suspension; ♀ in B moult end Dec.-early March, in mid March 80% in moderate or heavy moult. **(67,279).**

EIDER *Somateria mollissima*

P:10+1 r. S:-,D R:14

Ad postN complete, flight feathers simultaneous; ♂ B from mid June, wing mid July-late Aug., ♀ c1 month later. **Ad preN** partial (B), ♂ mid Aug.-Nov., ♀ Oct.-March. **PJ** and 1st preN partial (B, very variable) from Sept. in ♀ , Oct., in ♂ . 1st Imm postN like Ad but less complete and more prolonged (some B may be retained). **(111,126.261,279,541).**

COMMON SCOTER *Melanitta nigra*

P:10+1 r. S:-,D R:14

Ad postN complete, flight feathers simultaneous, flightless c3-4 weeks, ♂ mid July-mid Sept., ♀ Sept.-Oct.; R late March-early May, B from April suspended during P moult. **Ad preN** partial (B,R) slightly overlaps with postN B moult; ♂ Sept.-Dec. (mostly Oct.-Nov.), ♀ c1 month later. **PJ** partial (B, rarely T) from Sept. 1st Imm preN partial (B) from late Nov., finished April-May; R usually moulted April-May, but some already finished Jan. Flight feathers late June-Oct. of 2nd year. **(111,202,279).**

GOLDENEYE *Bucephala clangula*

P:10+1 r. S:-,D R:16

Ad postN complete, flight feathers simultaneous, flightless 3-4 weeks, ♂ mid July-mid Sept., ♀ c3 weeks later; B from June/early July, sometimes May in ♀ . **Ad preN** partial (B) starts when P and S full grown; ♂ in NP by late Oct., ♀ by early Dec.

PJ partial (B,R) in early autumn. 1st Imm preN partial (B, variable), mostly from Dec. 1st Imm postN like Ad but more prolonged and variable in timing, ♂ starts wing late June-late Aug, finishes mid July-late Sept., ♀ may be later. **(111,278).**

RED-BREASTED MERGANSER **Mergus serrator*

P:10+1 r. S:-,D R:18

Ad postN complete, flight feathers simultaneous, flightless c1 month, ♂ mid July-late Aug., ♀ c1 month later; B starts May in ♂, slightly later in ♀, finishing after wings. **Ad preN** partial (B,R), ♂ largely finished by Dec., ♀ c1 month later. **PJ** partial (B, variable), Oct.-Jan. 1st Imm preN partial (B,T,R variable), starts Dec., finishing in spring. **(111,279).**

GOOSANDER **M. merganser*

P:10+1 r. S:-,D R:18

Ad postN complete, flight feathers simultaneous, flightless c1 month, ♂ mid July-late Sept., ♀ a little later; part of B from mid June-Aug. but suspended during wing moult. **Ad preN** partial (B,T), overlaps with end of postN, most finish Dec., ♀ may be later. **PJ** partial (B,T,R) from Sept.; 1st Imm preN from Nov.; in South Dakota all R replaced by end Nov. **(8,111,157).**

RUDDY DUCK **Oxyura jamaicensis*

P:10+1 r. S:-,D R:18

Ad postN complete, flight feathers simultaneous, ♂ late July-Aug., ♀ from late Aug.; B after wings. **Ad preN** complete, flight feathers simultaneous March-April. **PJ** partial (B) from Aug. 1st Imm preN probably complete March-May. **(111,288, 536,570).**

ACCIPITRIFORMES AND FALCONIFORMES

Moult in Accipitriformes is complex, moult cycles in larger species taking more than one year to complete, in which case Primary moult is serially descendant but individual feathers may be omitted from a cycle resulting in a seemingly irregular moult pattern. Secondaries moult from 2 or 3 centres.

In Falconiformes moult of Primaries descendant and ascendant from P4 (exceptionally P5 or P6). For discussions on moult sequences see Brown and Amadon (1968), Mebs (1960), Miller (1941) and Stresemann and Stresemann (1960). For pterylosis see Compton (1938).

ACCIPITRIDAE

HONEY BUZZARD **Pernis apivorus*

P:10+1 v. S:13-14,D R:12

Ad postN complete, starting in breeding quarters, usually P1 and P2 replaced late July-early Sept.; moult finished in winter quarters. **PJ:** B and some WC moulted from beginning 2nd year, P from June. **(111,581).**

RED KITE **Milvus milvus*

P:10+1 v. S:-,D R:12

Ad postN complete, May (rarely April) — autumn, ♀ earlier than ♂; R start when inner P fallen, S and B later; median WC retained longer than GC and LWC. **PJ** partial (B, no WC) extent and timing uncertain; next moult complete at same time as Ad or earlier. **(111,602).**

WHITE-TAILED EAGLE **Haliaetus albicilla*

P:10+1 v. S:-,D R:12

Ad postN almost continuous but suspended Nov.-March; sequence of P and S seemingly irregular. **PJ** and subsequent, B begins May-June 2nd year, soon followed by R, P (descendant), and inner S descendantly from S16 or S17; outer S start ascendant moult from centres at S1 and S5, but moult at these centres sometimes postponed until 3rd year. Third generation of flight feathers appears at same centres usually 2 years later, fourth generation after a further 2 years or more, Juv P10 and last Juv S (usually S9 or S10) shed in 4th (occasionally 5th) year. Irregularities in basic sequence may occur with 2nd generation, more often later. **(111).**

MARSH HARRIER **Circus aeruginosus*

P:10+1 v. S:13?,D R:12

Ad postN complete April/May-Oct; ♀ starts with P1 at start of laying, ♂ later; R irregular, S from 3 centres, ascendant from

S1 and S5, descendant from S13. **PJ** starts in 1st winter with B and occasionally some R, wing and R at same time as Ad. **(111,500).**

HEN HARRIER **C. cyaneus*

P:10+1 v. S:13?,D R:12

Ad postN complete April/May-Oct., frequently longer; ♀ starts with P1 at start of egg laying, ♂ a few weeks later; R centrifugal, S from 3 centres (see Marsh Harrier); in USA ♂ moults more quickly than ♀. **PJ** like Montagu's Harrier? PostImm as Ad postN. **(111,517,520,521).**

MONTAGU'S HARRIER **C. pygargus*

P:10+1 v. S:13,D R:12

Ad postN complete, timing uncertain; some ♂ start in June, most in July, reaching P3-P7 before departure for winter quarters in Sept.; ♀ earlier, inner P shed rapidly in June; presumably suspend during migration and finish in winter quarters; R at about the same time as P, S from 3 centres (see Marsh Harrier). **PJ** some B in 1st winter (occasionally some R); moult of rest of B,R, and flight feathers presumably at same time as Ad; occasionally some Juv S retained until 3rd year. **(111,520).**

GOSHAWK **Accipiter gentilis*

P:10 S:-,D R:12

Ad postN complete April-Sept.; ♀ starts with P1 about start of laying, ♂ 15-30 days later; timing of P may vary but individual birds keep to same schedule each year; R1 dropped after P1, sequence variable; B starts June; S from 3 centres, ascendant from S1 and S5, descendant and ascendant from S13; moult of R and S may be suspended and finished early following year; duration of wing moult up to 171 days. **PJ** complete in 2nd year, starting later than Ad. **(111,423,514,588).**

SPARROWHAWK ***A. nisus*

P:10 S:14,D R:12

Ad postN complete May/June-Sept./Oct.; ♀ starts P moult during egg laying, ♂ c26 days later; some birds (both sexes) may suspend when feeding young (more ♀ than ♂), ♀ usually suspend at : P4/5/6, ♂ at P2/3/4; av P duration ♀ c110-130 days, ♂ c100-120

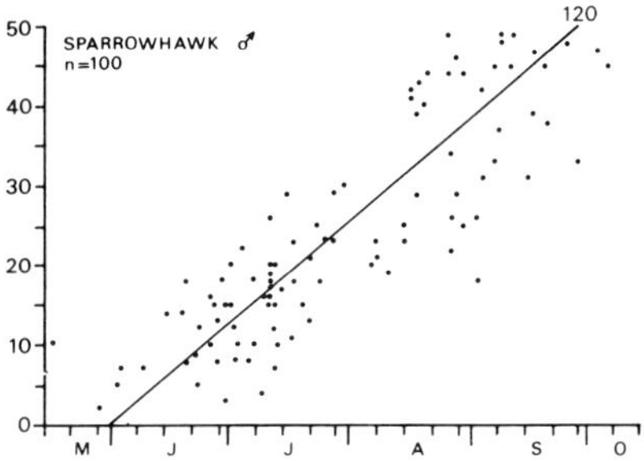

Much additional data from Newton and Marquiss (1982)

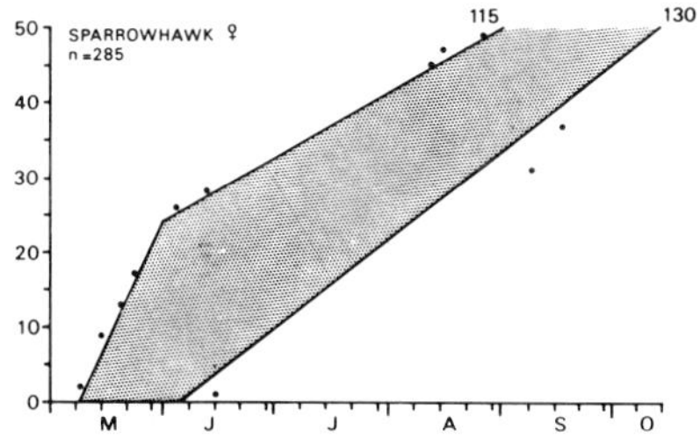

Much additional data from Newton and Marquiss (1982)

days; 1Y may take longer than older birds, variable depending on duration of suspension; start of S variable but most when P3/4/5 moulted, usually 3 centres at S1 and S5 (ascendant), and S10 (ascendant and descendant) but variable and moult may start at any feather in S tract; most start R when P3/4 moulted, av sequence 1-6-3-4-5-2, variable; B starts soon after P. **PJ** nearly complete in 2nd year, usually retain feather between tertials and scapulars until June 3rd year. **(111,338,410,423,537,588).**

BUZZARD **Buteo buteo*

P:10+1 v. S:13,D R:12

Ad postN complete March-Nov.; B from March/April; P from late April/May, ♀ starts during incubation, ♂ later; Ad ♂ 29 Sept. Pscore 23; S start at S5 usually after dropping P4 or P5; R from R1 at same time as wing, irregular and usually asymmetrical; P serially descendant, irregular and asymmetrical. **PJ** complete, B from Nov./Dec. of 1st year, wing and R at same time as Ad in 2nd year, P usually from early/mid May, P duration 158 in captive ♂; R from R1 symmetrical finishing with R5. 1st Ad postN complete but P10 may be retained, subsequently number of P moulted highly variable. Ad *vulpinus* start moult on breeding grounds then suspend, finishing in winter quarters. **(32,83,111, 459).**

GOLDEN EAGLE **Aquila chrysaetos*

P;10+1 v. S:17,D R;12

Ad postN complete, but not all flight feathers replaced each year, March/April-Sept.; P seemingly irregular; S from 3 centres at S1, S5, inner centre variable; GC usually in same sequence as S, median WC irregular; R irregular but often starts with R1; individual feathers may be omitted from one moult and not replaced until next cycle. **PJ** complete March/April-Aug./Sept., B and inner P (usually P1-P4/P5); suspends during winter, resumes from March of 3rd year, P descendant but P10 may be replaced before P9, P8-P10 may not be replaced until 4th year; R apparently replaced in 3rd year, irregular. **(111,286,530, 554,588).**

PANDIONIDAE

OSPREY **Pandion haliaetus*

P:10+1 v. S:20-21,D R:12

PJ starts early in 2nd year, P descendant, S ascendant from S1 and S5, R irregular: first cycle not finished until c5 years old. Next cycle starts before first finished, thus P serially descendant but time interval between cycles decreases with age resulting in seemingly irregular moult of P and S. Most active moult June/July-Aug./Sept., Oct./Nov.-Feb./March, suspends during migration. **(109,111).**

FALCONIDAE

KESTREL ***Falco tinnunculus*

P:10+1 v. S:13,D R:12

Ad postN complete; P ascendant and descendant from P4, with some individual variation in sequence; S start with S5 at same time as P, ascendant and descendant; other tracts later, UpperWC after 2 weeks, R after 21-23 days, UnderWC after 50 days; wing moult duration c130 days; B protracted lasting almost entire year. ♀ starts wing moult during incubation from mid May, finishes early Sept. (one Pscore 49 on 4 Nov.); ♂ starts c2 weeks later in early

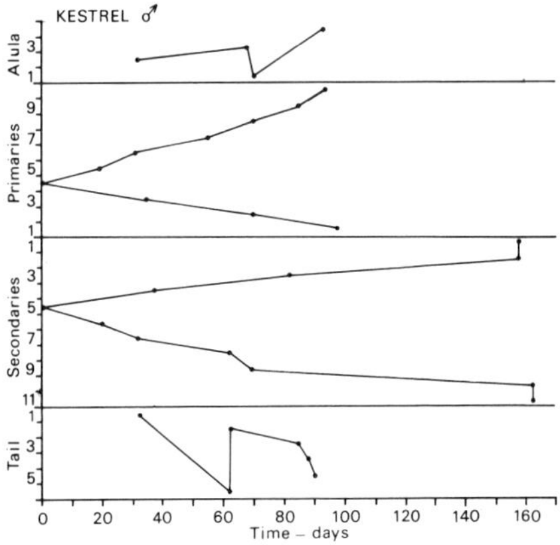

Sequence and timing of wing and tail moult in a captive male Kestrel. Redrawn from Piechocki (1956)

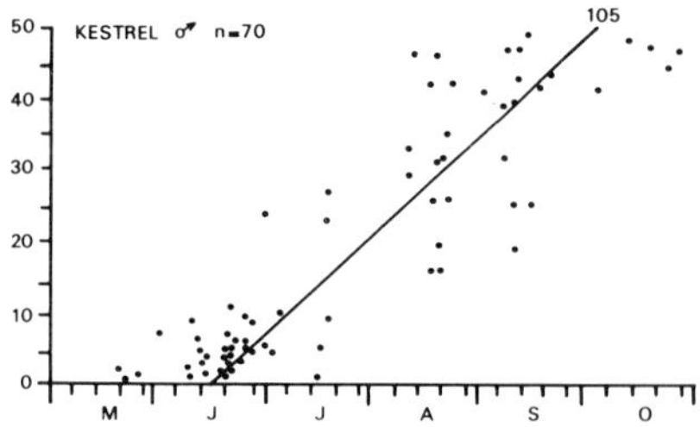

Much additional data from Village, Marquiss and Cook (1980)

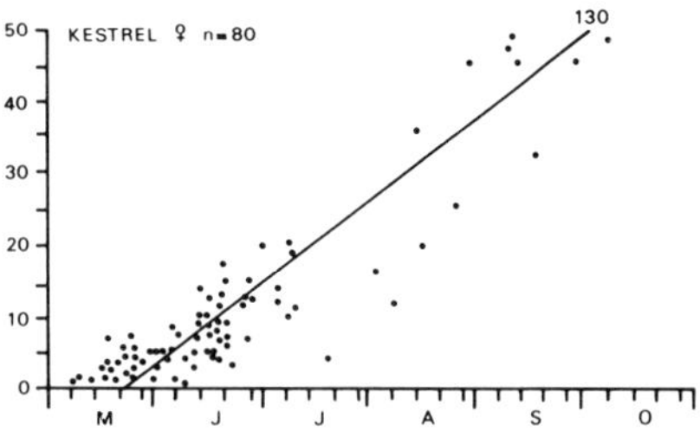

Much additional data from Village, Marquiss and Cook (1980)

June but apparently finishes at same time as ♀ thus moulting more quickly. **PJ** partial, B from soon after fledging to following summer, some WC occasionally replaced. Juv wing and R replaced June/July-Oct./Nov. of 2nd year. **(40,111,456,603, 662).**

MERLIN **F. columbarius*

P:10+1 v. S:14,D R;12

Ad postN complete June-Sept./Nov., ♀ somewhat earlier than ♂; P sequence variable 5-4-6 (or 4-5-6-, or 6-5-4)-3-7-8-2-9-1-10; S from S5 ascendant and descendant; R from R1 centrifugal, ending with R5. **PJ** partial Feb.-May, some B and R1 (occasionally more R); subsequent moult complete at same time as Ad. In UK ♀ 18 June Pscore 2, ♂ 16 Aug. Pscore 2, ♀ 8 Sept. Pscore 49. **(111,250).**

HOBBY **F. subbuteo*

P:10+1 v. S:11,D R:12

Ad postN complete Aug.-April; B starts Aug. and occasionally P4 or P4 and P5 replaced before migration; one Hampshire replacing P4 and P5 (Pscore 6) and S5 and S6 (both stage 4) in August, one ♀ Portugal 17 Aug. Pscore 13, ♂ Zambia 4 Jan. Pscore 36; moult finished in winter quarters; P sequence 4-5-3-6-7-2-8-9-10-1, duration P4-P10 138-149 days; S5 dropped shortly after P4, S continue after P finished. **PJ:** some start B in March/April but others not yet started by Aug.; P4 or P4 and P5 moulted July of 2nd year, finished in winter quarters. **(111,140,458,581,650).**

PEREGRINE **F. peregrinus*

P:10+1 v. S:13,D R:12

Ad postN complete April/May-Sept./Nov.; P sequence 4-5-3-6-7-2-8-1-9-10, ♀ starts with P4 usually after laying 3rd egg and may suspend when feeding young, ♂ starts when young hatched; 2+Y ♂ at Fair Isle replacing P4 and P5 (both stage 4) 6 July; R start 10-47 days after shedding P4 in sequence 1-2-3-4-5, finished before P; B moulted at same time as P; S sequence 5-6-7-4-8-3-9-2-10-11-1; moult finished in 128-185 days (captive birds). **PJ** complete starting with P4 in March of 2nd year, R start 6-25 days later, but much individual variation in timing; 2Y ♂ Wales 25 Aug. replacing P4 and P5 (Pscore 6). **(111,155,360,581).**

TETRAONIDAE

RED GROUSE **Lagopus lagopus*

P:10 S:15-16 (-18),E R:16

Ad postN complete; P1 falls in June, B starts Aug., finish Nov. but timing variable; UK Pscores: ♂ 20 July 31, ♂ 30 July 31, unsexed 19 Aug, 36, 43. Outer S begin with S3 ascendant, S1 and S2 moulted when S5 full grown, inner S simultaneous; R centripetal; R and wing moulted more quickly in ♀ than in ♂. **Ad preN** partial (most of B and inner S), ♂ May-June, ♀ March-May. **PJ** largely complete. P1 shed when c3 weeks old, P9 and P10 retained until complete moult in 2nd year. **(111,421,435).**

PTARMIGAN **L. mutus*

P:10 S:18-19,E R:16

Ad postN nearly complete, ♂ July-early Sept., ♀ c2 weeks later and retains more NP; moult starts with P1, B starts soon afterwards; R replaced quickly, centripetal, starting when P6 dropped; S ascendant from S3, S1 and S2 shed when ascendant series about ½ complete. Winter moult of pigmented B feathers continues to Dec. **Ad preN** partial (most B, inner S), ♂ May-June, ♀ quicker finishing in c1 month (May) with more extensive B moult than ♂. **PJ** partial (crown, hindneck, throat, most WC retained); P1 lost at c3 weeks old, P9 and P10 retained until complete postN moult in 2nd year. 4-5 weeks after start of PJ, only white feathers appear and Imm winter plumage acquired. **(111,253,504,678, 682).**

BLACK GROUSE **Tetrao tetrix*

P:10 S:-,E R:18

Ad postN complete, June-Oct./Nov. (occasionally starting May, finishing Dec.); ♀ with young later than ♂ or ♀ without young; R centripetal, almost simultaneous. **PJ** almost complete, starts when c15 days old, lasts 14-16 weeks; R almost simultaneous, starts when P5-P6 dropped at c7 weeks; P9, P10, S17 and S18 and some WC retained. Post Imm moult as Ad postN. **(111,244,581,605).**

CAPERCAILLIE **T. urogallus*

P:10 S:18-19,E R:18 (occ.20, exc.23,24)

Ad postN complete; ♂ starts with P1 in May at end of display period, ♀ when young hatch (late May/early June); S start at S15, followed soon afterwards by a second centre at S3; R centripetal; B July-Sept./Oct. **Ad preN** partial (head and neck) May-June. **PJ**

complete except for P9 and P10; starts when c18 days old, finishes in c3 months; R starts at c1 month finishing at c2 months; ♂ sheds P8 at c80 days, ♀ at c75 days. Post Imm like Ad postN but ♂ earlier, P1 shed in March of 2nd year. **(111,244,310,581).**

PHASIANIDAE

BOBWHITE **Colinus virginianus*

P:10 S:12,E R:10

Ad postN complete Aug.-Oct. (in USA), ♂ possibly earlier than ♀. **PJ** partial, starts with P1 at 26-30 days, P8 full grown by 150 days, P9 and P10 retained. **(497,620,679).**

RED-LEGGED PARTRIDGE **Alectoris rufa*

P:10 S:15,E R:14

Ad postN complete; starts with B and inner P late June-late Aug., R slightly later; B and R finish Oct., wing late Aug.-mid Nov. ♂ slightly earlier than ♀. **PJ** complete except for P9 and P10, starts with B and inner P when bird c½ grown; R shed with inner P, finish at same time as P8. **(111).**

PARTRIDGE ***Perdix perdix*

P:10 S:15,E R:14

Ad postN complete, P start mid June-early July (extremes late May, early Aug.), ♀ starts when eggs hatch, R centrifugal; S12-S15 dropped simultaneously at same time as inner P, when c½ P replaced S3 dropped and moult proceeds ascendantly to S11, S1 and S2 dropped when outer P completing growth; B and R finish mid Aug.-late Sept., outer P sometimes not until early Nov. **Ad preN** partial, ♂ head and neck early May-June, ♀ head and much B March-May. **PJ** complete except for P9 and P10; P1 lost when c3½ weeks old and before full Juv plumage attained (thus both inner and outer P may be found growing at same time on one wing), P8 dropped at 12 weeks and full grown at 16-17 weeks (early Sept.-late Nov.); B and R start at 5-6 weeks, S start with S3 at 6-7 weeks. **(111,581).**

QUAIL **Coturnix coturnix*

P:10 S:14,E R:10-12

Ad postN complete, starts with B and inner P soon after eggs hatch, early June-mid Aug.; finished late July-late Sept. but some late birds suspend and start migration with worn outer P, active flight feather moult during migration exceptional; R centrifugal, starts during growth of P4 or P5; B finished in winter quarters. **Ad preN** partial (B) Feb.-April/May. **PJ** nearly complete but outer 5-6 P retained; B starts at 17-21 days old, P1 drops when 22 days old; most B finished at 6 weeks, P1-P4/P5 replaced by c7 weeks old; moult suspended during migration. 1st preN Sept.-Nov., usually in winter quarters; P moult continues to P7, R replaced and B moulted; head may be moulted again in March. **(111,581,593).**

PHEASANT ***Phasianus colchicus*

P:10 S:16,E R:18 (16-17 fre. ♀, exc. ♂)

Ad postN complete; ♂ starts June, ♀ later, after hatching of young; R centripetal; total duration c3 months, outer P and centre R last to finish, Sept.-early Oct. **PJ** complete (all P replaced); Juv plumage appears from c10 days, PJ moult starts with alula when c3 weeks old, followed by P1 and S3 at 3-4 weeks; generally in 1W plumage when 4½ months old (Sept.-Nov.) but outer P and centre R, some T and UnderWC not quite finished then. **(93,111,604,605,686).**

GOLDEN PHEASANT **Chrysolophus pictus*

P:10 S:-,E R:18

In captive birds **Ad postN** complete; ♂ starts with head, R and inner P mid May-early June, finishes Sept. but outer P not until Nov. in some; ♀ starts some weeks later. **PJ** complete (all P replaced); in complete 1st Ad plumage by Oct./late Nov.; some Ad NP feathers may appear in spring 2nd year; thereafter moult as Ad. **(111).**

LADY AMHERST'S PHEASANT *C. amherstiae* as Golden Pheasant.

RALLIDAE

WATER RAIL **Rallus aquaticus*

P:10 S:-,E R:12

Ad postN complete, flight feathers simultaneous, flightless c3 weeks, early July-early Sept; ♂ Minsmere 8 Aug. all P and S stage 1; B starts soon after breeding (July, sometimes June), usually finished by late Aug./early Sept. (Nov.). **Ad preN**

partial (head, some B) Feb.-May. **PJ** partial (B, sometimes R), starts soon after fledging, finish mid July-early Oct. (Dec.). 1st preN partial (B), very variable in timing and extent. **(41,111,581).**

SPOTTED CRAKE **Porzana porzana*

P:10 S:-,E R:12

Ad postN complete, flight feathers simultaneous, regrown in c3 weeks; B starts when flight feathers replaced; moult usually in or near breeding area July-Oct.; starts autumn migration with fresh NNP from Sept. **Ad preN** partial (B), Feb.-March. **PJ** partial (head and B) from July (most late Aug.-early Sept.), finish mid Sept.-mid Oct.; often moults during stops on migration. **(85,111,581,600,621).**

CORNCRAKE **Crex crex*

P:10 S:13-14,E R:12

Ad postN complete, flight feathers simultaneous; starts shortly after nesting mid July-late Aug., finished late Aug-mid Sept., sometimes B not finished until Oct.; some may postpone moult until after post-nesting dispersal or until reaching winter quarters. **Ad preN** partial (B,R and possibly some WC) Dec.-March. **PJ** partial (B) starting when flight feathers not quite full-grown (mid July-mid Sept.), finished in c1 month. 1st preN partial (B, some WC) Feb April. **(111).**

MOORHEN ***Gallinula chloropus*

P:10+1 v. S:12-14,E R:12

Ad postN complete, flight feathers simultaneous; B starts late May-late June; flight feathers late June-late Aug., P grow at c5.0mm/day in South Africa; most finish B by late Sept. but a few not until Nov. **Ad preN** partial (B), not in all birds, March-June. **PJ** partial (B) starts at 15-18 weeks old, mainly finished Sept. 1st preN partial (B) very variable in timing and extent. **(40,111,165,199,259,293,581,713).**

COOT ***Fulica atra*

P:10+1 v. S:17-20,E R:14 (12-16)

Ad postN complete, flight feathers and all WC simultaneous; starts with B from May but slow during nesting; non breeders moult wing from mid June, later in breeding birds (late June-early Sept.) but very variable, ♂ slightly earlier than ♀; R lost soon after flight feathers. **Ad preN** partial (head neck) Dec.-May. **PJ** partial (B), starts before wings full grown, mainly finished before Oct., some not until Dec. 1st preN partial (B) Dec.-May. **(111,174,314,581).**

CHARADRIIDAE AND SCOLOPACIDAE

Waders typically shed P1-P5/P6 more or less simultaneously, P6/P7 is then dropped when the inner Primaries are nearly full-grown and the rest of moult is more leisurely. In those Palaearctic species with a wide latitudinal range in the winter quarters the timing and duration of moult may vary markedly, birds spending the 'winter' non-breeding season in the tropics and south of the equator tend to take longer to moult. This has been documented for several species e.g. Sanderling *Calidris alba*, Little Stint *C. minuta* and Curlew Sandpiper *C. ferruginea*, and is likely to occur in others. Generally non-breeding (Immature?) birds start to moult before breeding adults, especially if non-breeders remain in the 'winter' quarters. First-summer birds of a number of species have a partial moult of the outer primaries in the 'spring', followed by a complete moult starting in late 'summer', this being most frequent in tropical areas. There are few accounts of the partial Pre-Breeding (Nuptial) moult of most waders, and although there is much information on the plumage status of birds collected, this is widely scattered and not readily available. For general accounts of moult in waders see e.g. Anon (1979), Hale (1980), Minton (1976), Khrokov (1978a), Prater *et al.* (1977), and Stresemann (1963c).

HAEMATOPODIDAE

OYSTERCATCHER

****Haematopus ostralegus*

P:10+1 m.r. S:18-20,D R:12

Ad postN complete; Wales postN Ads start July, on Wash c5% in suspended P moult on arrival; S1 drops with P6/P7; R1 drops with P1, irregular; P duration c100/c135 days

Iceland, 130-140 days Waddenzee, c90-105 days Wales, c90 days Wash. **Ad preN** partial (B, centre R, inner S) Jan.-May. **PJ** partial (B, inner S, may include centre R) Aug.-Dec. On Wash 1S birds start complete moult late April, P duration c135 days; 2S birds start late May, P duration c120 days. **(3,76,116,220,263,274,385,466,707,710).**

RECURVIROSTRIDAE

AVOCET **Recurvirostra avosetta*

P:10+1 m.r. S:-,D R:12

Ad postN complete June-Jan. (2 on Waddenzee in Oct./Nov. had finished); one ♀ Holland Pscore 1 whilst incubating. **Ad preN** partial (B, centre R or more, inner S) Jan.-June. **PJ** partial (B, some or all R, inner S, some WC) Aug.-Jan. **(76,126,274, 467,672,710).**

BURHINIDAE

STONE CURLEW **Burhinus oedicnemus*

P:10+1 m.r. S:-,D R:12 (14)

Ad postN complete from May, finishing Oct./Nov. (exceptionally Dec.). P moult apparently slow, 1 in May with inner 3 P growing but in later stages of moult only 1 P growing at a time in each wing. UK Pscores: 29 Aug. 44, 4 Sept. 42, 25 Sept. 49, 28 Sept. 39, 1 Oct. 34, 11 Oct. 49. **Ad preN** partial (B,R, inner S, some WC) March-May. **PJ** partial (B, some R, some WC) Sept./Oct. **(274,710).**

CHARADRIIDAE

LITTLE RINGED PLOVER ***Charadrius dubius*

P:10+1 m.r. S:-,D R:12

Ad postN complete July-Nov.; some (but probably not all) British breeding birds start wing moult before migrating, UK Pscores: 2 July 0; 5 July 4; 10 July 7; 13 July 5; 27 July 2; 29 July 24; 1 Aug. 0,2; 5 Aug. 27; 11 Aug. 18; 19 Aug. 14; **Ad preN** partial (B, inner S, occasionally R, some WC) Feb.-May. **PJ** partial (B,R, inner S, some WC), autumn. **(194,274,316,467,487,710).**

RINGED PLOVER ****C. hiaticula*

P:10+1 m.r. S:-,D R:12

Ad postN complete, R start shortly after P, S start at Pscore 25-35. UK breeding birds start July/Aug., some before breeding complete. In Greenland 18% of B feather mass is replaced on breeding grounds. Some Icelandic *hiaticula* and USSR *tundrae* start P moult before migration, on Wash 10% in suspended P moult on arrival, in Morocco c30% show suspended moult; in Morecambe Bay moult well underway by early Aug., nearly finished early Oct. P duration c100-104 days for 2 retraps in Morocco. **Ad preN** partial (B, occasionally some R, some T and WC) March-May. **PJ** partial (B,R,T, some WC) Aug.-Jan., extent of P moult variable but renewal of all P probably unusual except in tropical Africa where complete moult Dec./Jan.-March; in UK B moult extensive and some 1W become indistinguishable from Ad. **(3,53,76,100, 133,167,194,274,316,376,386,443,467,577,581, 594,710).**

KENTISH PLOVER **C. alexandrinus*

P:10+1 m.r. S:-,D R:12

Ad postN complete, July-Sept. (Nov). In Holland, Morocco and Iraq some may start P moult whilst breeding. Main moulting area(s) for European birds unknown. 3 retraps in Morocco give P duration 90,95,100 days. **Ad preN** partial (B, occasionally R, some T and WC) March-May the extent of Ad preN moult requires further study, Ad ♂ in Morocco retained much rufous on head through postN moult. **PJ** partial (B,R,T, some WC) Sept.-Dec. **(133,194,274,291,316,386,463,467,581,710).**

DOTTEREL **C. morinellus*

P:10+1 m.r. S:-,D R:12

Ad postN complete, starts on or near breeding grounds July/Aug., by early Sept. inner P and c½ B replaced, suspends during migration (some may start migration with growing P) and finishes in winter quarters Dec./Jan.-end Feb./early March. **Ad preN** partial (B, some T and WC, sometimes centre R), progresses slowly March-June. **PJ** partial (B, some T and WC, sometimes centre R) Sept.-Nov., before/during migration. **(126,274,316,482,710).**

GOLDEN PLOVER **Pluvialis apricaria*

P:10+1 m.r. S:-,D R:12

Ad postN complete July-Nov. In Iceland moult starts before breeding finished and lasts c4$\frac{1}{4}$ months. 1 Ad caught on eggs in Wales 6 June Pscore 7. In Waddenzee Pscores 14-34 in Aug. **Ad preN** partial (B, some T and WC, occasionally centre R) Feb.-May. **PJ** partial (B,T, some WC) Sept.-early Nov. **(76,220,274,316,581,710).**

GREY PLOVER ***P. squatarola*

P:10+1 m.r. S:-,D R:12

Ad postN usually complete July-Dec. Ads arriving on Wash in Aug., 25-40% in suspended P moult, moult starts/resumes in Aug. (some not until early/mid Sept.); 16% of Ads between late Nov. and end March retain P10 or P9 and P10, most apparently resume in mid March but 4 of 10 had not resumed by end March. P duration 90-100 days; c30% in suspended moult in

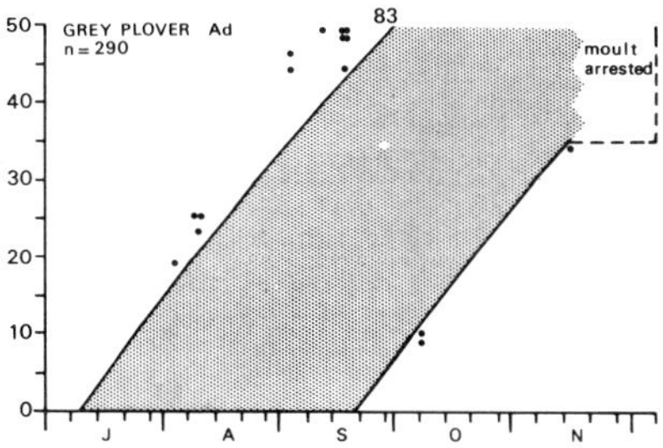

Redrawn from Branson and Minton (1976)

Morocco. **Ad preN** partial (B,R,T, some WC) Feb.-May. **PJ** partial (B, some T and WC, occasionally R) Sept.-Nov. (Jan.), 1S complete moult earlier than Ads, finish late Aug. on Wash. **(3,53,76,79,133,274,316, 333,364,386,467,484,548,581,710).**

LAPWING ****Vanellus vanellus*

P:10+1 m.r. S:-,D R:12

Ad postN complete, early June-Oct. (Nov.); S1 drops with P7; R starts when P6 growing, finishes when P8 or P9 finishing, sequence variable but most 1-3-4-2-5, R6 may be first or last. P duration c114 days **(12).** Some migrate to UK from Europe in active wing moult. **Ad preN** partial (head, some nape, throat, breast, some scapulars and WC, occasionally some T) Feb.-May. **PJ** partial (B,T, some WC, exceptionally some R), July-Dec. **(12,115,198,377,413, 484,545,581,710).**

SCOLOPACIDAE

KNOT ***Calidris canutus*

P:10+1 m.r. S:15,D R:12

Ad postN complete July-Nov.; on Wash c1% arrive in suspended P moult; P duration 90-100 days (Waddenzee); S start at Pscore 20-30 and finish before P; R start at Pscore 5-30. **Ad preN** partial (B,T, some WC, sometimes centre R) Feb.-June. **PJ** partial (B, T, centre R, some WC) Sept.-Dec. 1S moult P late May-Aug./Sept. in Europe. **(3,53,75,76,133,274,334,376,386, 467,481,482,581,615,657,658,710).**

SANDERLING ***C. alba*

P:10+1 m.r. S:-,D R:12

Ad postN complete July/Aug.-Oct./Nov., in South Africa start on arrival and finish mid Feb. In Greenland 35% of B feather mass is moulted on breeding grounds. S start before Pscore 20. No evidence P suspension in Europe or South Africa, but c70% in Surinam had moult suspended in at least one feather tract. **Ad preN** partial (B, some T and WC, occasionally centre R) March-May. **PJ** partial (B, some T and WC,R) Sept.-Dec., in East Africa some moult outer P in 1W. **(53,76,133,167,274, 376,467,482,549,566,581,673,710).**

LITTLE STINT **C. minuta*

P:10+1 m.r. S:-,D R:12

Ad postN complete Aug.-March, timing and duration variable; P duration 50-60 days Morocco, 100-115 days South Africa in East Africa most c130 days, some < 100 others > 150 days. **Ad preN** partial (B,R,T, some WC) March-May in East Africa 5-10% also renew some (all) P. **PJ** variable in extent and timing, in East Africa most have complete moult Dec./Feb.-March/April when P may be replaced in 60-80 days, others only replace outer P and B,T and some WC. Early April Saudi Arabia 15 of 20 were in active P moult (scores 31-48, av 38: D. M. Francis pers. comm.). **(53,76,119, 133, 144, 274, 369, 386, 443, 467, 482, 558, 581,631,632,710).**

TEMMINCK'S STINT *C. temminckii*

P:10+1 m.r. S:-,D R:12

Ad postN complete, starts July-Sept, then usually suspended and finished Feb.-April, a few finish Oct./Nov. **Ad preN** partial (B, centre R, some T and WC?) Feb.-May. **PJ** partial (B,T, some WC) from Aug. (do not start before leaving USSR), outer P moulted Dec.-May. **(126,274,467,482, 710).**

CURLEW SANDPIPER *C. ferruginea*

P:10+1 m.r. S:10,D R:12

Ad postN complete, Sept.-Feb. (B from mid July), very few in wing moult in West Europe. Timing and duration P moult variable; ♂ moult before ♀ in Morocco and Mauritania but not in South Africa of Tasmania; P duration 40+ days Morocco (one retrap 40 days), 90-100 days South Africa, 125-130 days Tasmania. In West Africa 1.6-5.3% in suspended P moult in autumn. **Ad preN** partial (B, centre R, T, some WC?) Feb.-May. **PJ** partial Aug.-Dec., in tropics moult some (outer) or all P, and T from Feb. **(53,76,120,133,153,274, 275, 366, 378, 386, 436, 443, 467, 484, 581, 618,632,673,706).**

PURPLE SANDPIPER **C. maritima*

P:10+1 m.r. S:-,D R:12

Ad postN complete, July-Sept.; S1 dropped when Pscore 15-20, S2 at Pscore 30, remaining S rapidly (Iceland); R basically centrifugal but variable, may start early in P moult but most not until Pscore c30. Iceland ♀ starts c7 July, ♂ c15 July; Spitsbergen start late July and probably finish before migration; in Scotland P probably start in July soon after birds arrive and c½ population finished by end Sept. P duration 40-50 days in Iceland. **Ad preN** partial (B, sometimes some T) in spring. **PJ** partial (B); T and most WC retained. **(20,58, 220,274,385,581,710).**

DUNLIN ****C. alpina*

P:10+1 m.r. S:15,D R:12

Ad postN complete, moult variable according to race: *alpina* may start P before migration and suspend, *arctica* and *schinzii* usually migrate before moulting (in Alaska complete moult on or near breeding grounds before migration); S start when

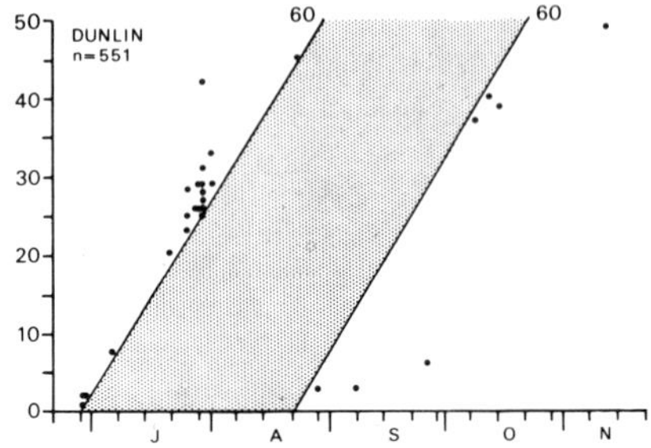

Pscore 20-30, R start at same time as P. On Wash 2% *alpina* arrive in suspended P moult, birds in active P moult from late June-Oct., P duration 59-89 days, late starting birds moulting more quickly. On Waddenzee most moult late June-late Sept., P duration 87-94 days, ♀ moulting more quickly than ♂. In NW Africa moult mid June-end Nov., P duration probably 60-70 days but some birds migrate whilst in active wing moult. **Ad preN** partial (B, some T) March-June variable; on Severn Estuary *alpina* undergoes whole preN moult, but *arctica* and *schinzii* arrive in almost FNP. **PJ** partial (B, some T and WC) Aug.-Nov., B moult c50 days behind that of Ads at Camargue. 1st preN moult partial as Ad (at least in *alpina*). **(3,75,76,77,167,168,181, 257, 258, 274, 282, 328, 377, 386, 411, 428, 464,467,484,556,565,581).**

RUFF ***Philomachus pugnax*

P:10+1 m.r. S:15,D R:12

Ad postN complete, timing rather variable; formerly thought that ♂ moulted before migration and ♀ after, but this is not usual, however ♂ may moult ahead of ♀, 3 weeks earlier in Kenya. Differences in timing between sexes complicated due to mating system and fact that ♀ tend to winter further South than ♂. P duration minimum 55 days in UK, 50-60 days Germany, 96 days South Africa, in Kenya from Pscore10/20-50 takes 110-130 days. 18% in Kenya in suspended P moult. **Ad preN** partial (B,R, some T and WC) Jan.-April, starting in winter quarters, ♂ earlier than ♀ in Kenya. **PJ** partial (B, some T and WC) Aug.-Dec., some moult outer 2-4 P Jan.-April. **(9,53,145,195,274,346,386,432,443,446,467, 516,522,580,581,632,710).**

JACK SNIPE **Lymnocryptes minimus*

P:10+1 m.r. S:-,D R:12

Ad postN complete July-Oct. before migration, but one Lancashire 17 Oct. Pscore 39. **Ad preN** partial (B,R, some T and WC) from Jan./Feb, moult at peak in mid April in USSr. In UK 3 birds in preN moult: 2 March S14 stage 3, 5 April R1 and R2 stage 2, 12 April R1 stage 3. **PJ** partial, may be finished during autumn migration. **(274,482,635,710).**

COMMON SNIPE ****Gallinago gallinago*

P:10+1 m.r. R: 14 or 16 (12-18)
(exc. full length) S:15-16,D

Ad postN complete June-late Oct.; in UK one in suspended/arrested moult having replaced P1-P8 10 March; in North America may begin during incubation, otherwise soon afterwards, ♀ c3 weeks later than ♂; in Germany many arrive having replaced P1-P3/P5 and suspended, most

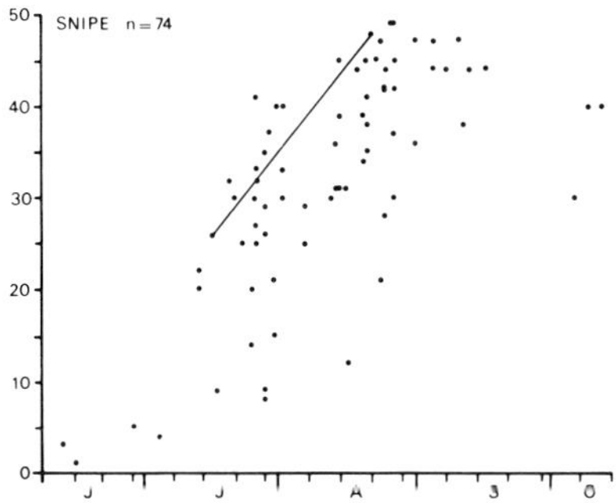

then finish but a few suspend again after partial moult and presumably finish elsewhere; R in quick succession, centrifugal; S start when c⅔P replaced, when S1 full grown S2-S10 (or S11) shed simultaneously; Upper WC renewed before S, moult of Under WC does not start until S fully grown in Germany (not so regular in USA?). **Ad preN** partial (B,T, some/all R, most WC) Jan.-May. **PJ** partial (B,R,T, some WC) July-Dec./Jan. 1st preN partial (B,T, some WC, some/all R) Jan.-early May in USA. **(56,274,419,482,635,697,710).**

WOODCOCK **Scolopax rusticola*

P:10+1 m.r. S:-,D R:12

Ad postN complete late June/July-Sept./Nov., ♂ earlier than ♀; UK Pscores: 25 Aug. 29, 9 Sept. 41, 2 Oct. 40, 15 Nov. 50, 22 Nov. 49. **Ad preN** partial (head, some B,T and WC, sometimes some R), Feb.-May. **PJ** partial (B,R,T, most WC) autumn. **(104,126,274,532,710).**

BLACK-TAILED GODWIT **Limosa limosa*

P:10+1 m.r. S:-,D R:12

Ad postN complete June-Oct.; UK Pscores: 20 July 12, 9 Sept. 17, 11 Sept. 26; in Netherlands P1-P6/P7 moulted late May-July/Aug., then suspend, possibly finishing in Morocco; P duration est c100 days Morocco. **Ad preN** partial (B, some T and WC, sometimes R) Feb.-June, ending after birds have started nesting. **PJ** partial (B,T, part/all R, some WC) Aug.-Jan. 1S start complete moult May/June. **(126,274,305, 383,386,467,482,640,710).**

BAR-TAILED GODWIT ***L. lapponica*

P:10+1 m.r. S:- D R:12
(exc. 11)

Ad postN complete Aug.-Nov./Dec.; P duration 90-100 days in Mauritania and UK (up to 130 days on Moray Firth), in Netherlands ♂ 100 days, ♀ 116 days, ♀ starting somewhat earlier; S start when Pscore 20-30, ascendant and descendant; on Wash c1% in suspended P moult on arrival. **Ad preN** partial (B, some R,T and WC) Feb.-May/June. **PJ** partial (B, usually some R, T and WC) Sept.-Jan. 1S start complete moult from May. **(3,75,76,133, 203,274,482,484,609,710).**

WHIMBREL **Numenius phaeopus*

P:10+1 m.r. S:19,D R:12

Ad postN complete, B starts on breeding grounds, P start on wintering grounds Aug./Sept., finish Nov./Jan. **Ad preN** partial (B,R, some T and WC) Feb.-May. **PJ** partial (B, some R, T and WC) Aug.-Feb. 1S variable, some start Jan./March, others not until May/June. **(53,76,133, 274,275,467,482,484,710).**

CURLEW ****N. arquata*

P:10+1 m.r. S:20-21,D R:12

Ad postN complete, late June-Nov.; P1-P3 (sometimes P4/P5) dropped more or less simultaneously, together with S18-S21.

then P and S in sequence: S17-P4-S16-P5-S1-P6-S2-S3-S4-S15-S5-P7-S6/7/8-S9-P8 S10-S14-S11-P9-S12-S13-P10. P duration c100 days Wash, 70-80 days Germany, 100 days Waddenzee; R starts when P1 dropped. **Ad preN** partial (B,R, some T and WC) Feb.-May. **PJ** partial (B, some WC, sometimes centre R and T) Aug.-Dec. (Feb./March). 1S start complete moult May/June, probably ending Aug./Sept. **(3,39,76,274,467,478,482,501,502,710).**

SPOTTED REDSHANK
**Tringa erythropus*

P:10+1 m.r. S:-,D R:12

Ad postN complete, late July-Oct.; in Norfolk on 27 July av. Pscore 7 Ad. 5.9,53 1S av. Pscore 33; on Waddenzee all birds FNP mid June, by mid July intermediate plumage, B finished by end Aug., on 23 July some (Ad?) starting P, other (1S?) nearly finished P: of 6 Ads in Morocco in Sept., one had suspended with 8 new P, 2 had renewed all P, others had Pscores 30,24,24. **Ad preN** partial (B,T, most R, some WC) March-May. **PJ** partial (B, some R,T, and WC) Aug.-Feb., most do not start before leaving breeding grounds. **(76,102,274,275, 386,482,484,710).**

REDSHANK ****T. totanus*

P:10+1 m.r. S:-,D R:12

Ad postN complete, timing variable June-late Oct./Nov., according to locality and to race; S start when Pscore 25-30, R often finished after P; British breeding *britannica* moult earlier than Icelandic *robusta;* on Wash 1% arrive in suspended P moult; P duration variable: estimates of 70/74-106 days for one Scottish sample depending on method of analysis, Waddenzee 125 days, Morocco 80-100 days; on Waddenzee *robusta* undergo complete moult late June-Nov., while some European *totanus* suspend P shortly after starting and migrate, others migrate whilst in active wing moult, probably making short flights between short-term staging areas. **Ad preN** partial (B, sometimes R, some T and WC) Jan.-May, extent of Ad preN variable throughout range, those in UK retain more NNP than elsewhere. **PJ** partial (B, usually some/all R, T and WC) Aug.-Jan. 1S may start P from April. **(3,75,76,127,133,182, 219, 274, 335, 386, 467, 482, 596, 598, 599, 611,710).**

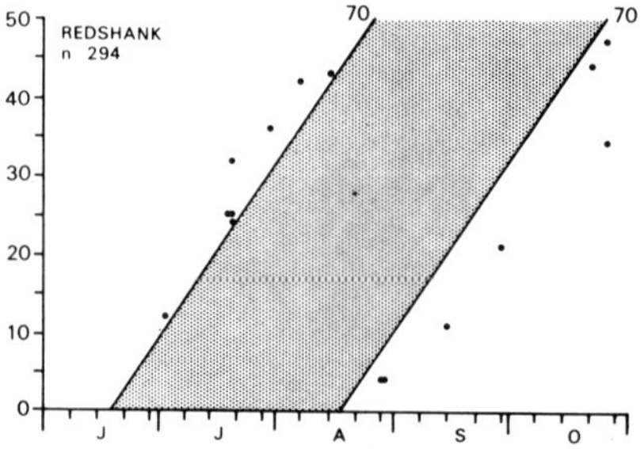

GREENSHANK ***T. nebularia*

P:10+1 m.r. S:17,D R:12

Ad postN complete, start June/July; on Waddenzee many still not started in Sept. while others suspended, most in East Africa probably arrive in suspended P moult, and at least some in South Africa; P duration c75 days (one retrap in Morocco); timing of end of moult variable, late Nov.-early Feb. in East and South Africa. **Ad preN** partial (B, some R,T and WC) Jan.-May. **PJ** partial (B, some R,T, and WC) Aug.-March, in East and South Africa outer 2-4 P may be replaced from Jan; 1S complete from late May. **(53,76,274,386,432,443,467, 482,484,632,633,710).**

GREEN SANDPIPER ***T. ochropus*

P:10+1 m.r. S:-,D R:12

Ad postN complete starting (late June) late July/early Aug., some finish before migrating, others suspend and finish in

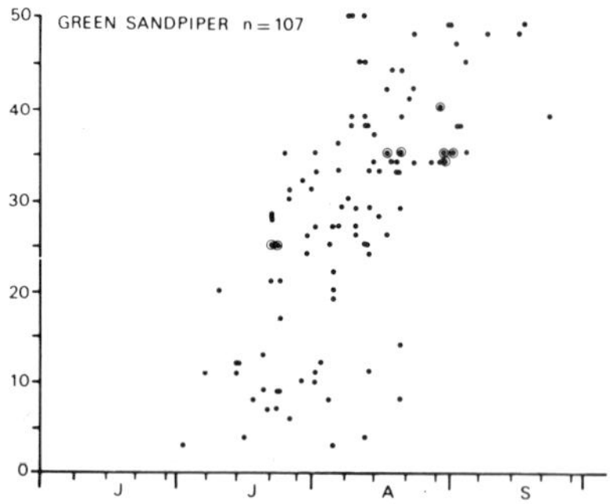

Circled scores were of birds which had arrested or appeared to be arresting their primary moult.

winter quarters by Nov./Dec., some migrate whilst in active P moult (UK, Iran). **Ad preN** partial (B,R, some T and WC) Dec.-May. **PJ** partial (B,R, some T and WC) Aug.-Dec. **(274,312,377,443,467,482, 484,710).**

WOOD SANDPIPER **T. glareola*

P:10+1 m.r. S:15,D R:12

Ad postN complete, Scandinavian birds moult in South Europe from late July-Oct.; one (1S?) Norfolk 26 July Pscore 23; most suspend and finish in African winter quarters up to Jan./Feb., in Camargue only 7 of 42 had finished P moult Aug./Sept., remainder had suspended after replacing

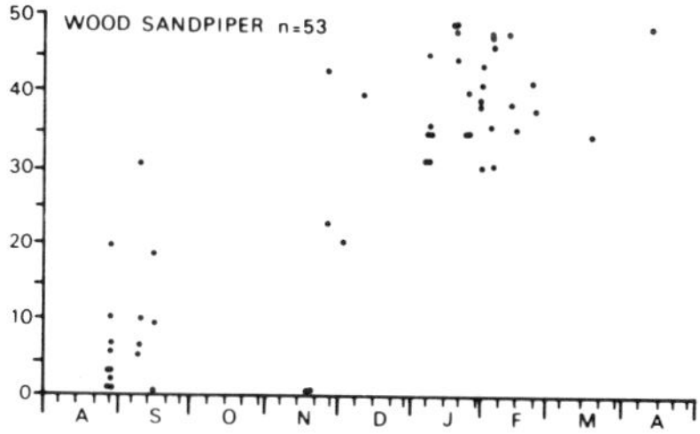

Data from the Lusaka area of Zambia (1972-75) provided by D. M. Francis

P1-P6/P7. P finished Dec./April in Zambia (D. M. Francis pers. comm.), P duration c3½-4 months South Africa. **Ad preN** partial (B, some R,T, and WC) Jan.-May. **PJ** partial (B, some R,T, and WC) Aug.-Jan., many (40-50% in East Africa) replace outer 3-5 P from Jan.-May. **(53, 76,255,274, 432,443,482,581,632,650,710).**

COMMON SANDPIPER **Actitis hypoleucos*

P:10+1 m.r. S:15,D R:12

Ad postN complete, many start Aug./Sept. and suspend, finishing up to March; most leave UK before starting, only c3% in active moult, none suspended; P duration c3-4 months in East and South Africa. **Ad preN** partial (B,R,T, some WC) Feb.-May. **PJ** partial (B,R,T, some WC) Aug.-Nov., birds aged as Juv P moult in Sept., Morocco; in Africa 1W moult outer or all P Dec./Jan.-early April, P moult more rapid than in Ad, duration c2 months; S moult rapid and irregular. **(53,76,88,274,386,443,445,467, 482,632,710).**

TURNSTONE ***Arenaria interpres*

P:10+1 m.r. S:-,D R:12

Ad postN complete July-Oct./Nov. (later in south of range); 15% of B feather mass is replaced on breeding grounds in Greenland; on Wash most start P moult 1-20 Aug. and finish late Oct., in Morecambe Bay starts at similar time but apparently continues later; 1% arrive on Wash in suspended P moult; S start when Pscore 15-30, finished before P; P duration c70 days Iceland, c70-75 days

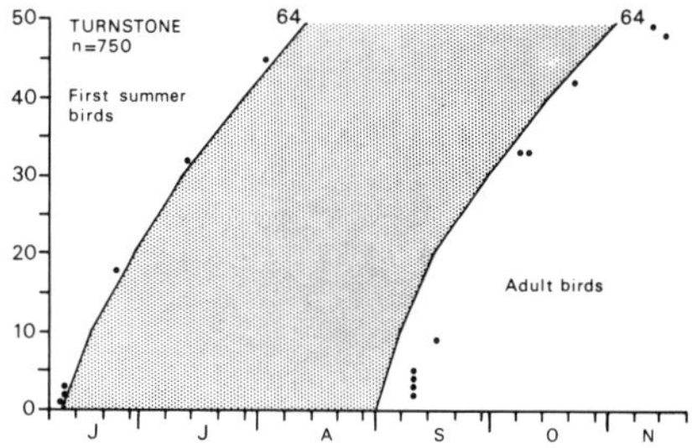

The left-hand side of the stippled area covers the moult of first summer birds (n=150, mean start second half of June) and the right-hand side of adults (n=600, mean start early August). Data from the Wash redrawn from Branson, Ponting and Minton (1979).

Waddenzee, c60 days Morocco. **Ad preN** partial (B,R,T, some WC) Feb.-June. **PJ** partial (B,R,T, some WC) Aug.-Nov., on Waddenzee 11 of 95 had replaced centre R in Sept., in south a few may moult inner P and arrest. 1S complete moult June-Aug./Sept. **(3,53,75,76,80,101,133,167,274, 283, 284, 333, 385, 386, 467, 482, 548, 595, 597, 710).**

RED-NECKED PHALAROPE **Phalaropus lobatus*

P:10+1 m.r. S:-,D R:12

Ad postN complete, timing variable; in Finland some ♂ start B while incubating, others, as ♀, not until they leave breeding grounds; in other areas both sexes start B on breeding grounds; P usually moulted in winter quarters from Nov., but some ♀? may moult July-Sept. **Ad preN** partial (B,T) Feb.-June. **PJ** partial, P moult recorded Feb.-June by **(581)**, but not by **(482)**. **(126,160,254,256,274,280,482,484,581,710).**

GREY PHALAROPE *P. fulicarius*

P:10+1 m.r. S:-,D R:12

Ad postN complete, B starts from early/mid July, some moult P near breeding grounds from Aug./Nov., others apparently later. **Ad preN** partial (B,R,T, and probably some WC) March-May. **PJ** partial (B, a few WC) in autumn, some 1S may moult outer P? **(126,220,274,311, 482,710).**

STERCORARIIDAE

ARCTIC SKUA *Stercorarius parasiticus*

P:10+1 m. S:-,E R:12

Ad postN complete, B starts Sept. (end June?), flight feathers in winter quarters Oct.-April. **Ad preN** Jan.-April (B). **PJ** complete? Nov.-March, P still moulting June. **(126,581,710).**

GREAT SKUA *S. skua*

P:10+1 m. S:-,E R:12

Ad postN complete, B June-Sept., flight feathers moult slowly in winter quarters Aug.-Jan./March, 1 or 2 P growing at one time, P duration 150-180 days; UK Pscores: 30 Jan. 45, 12 Feb. 46, 25 March 48. (In contrast the Southern Skua *S. maccormicki* moults rapidly late May-early Aug., P duration c45 days). **PJ** not studied. **(126,511,581,710).**

LARIDAE

Moult patterns are generally uncomplicated (Dwight 1901, 1925, Stresemann 1963c). The timing of a particular 'age-specific' moult may be difficult to determine due to great variation in the plumage of birds of the same age (Monaghan and Duncan 1979, Poor 1946). It is important to remember this when considering moult in birds of unknown age (i.e. not marked as pulli).

MEDITERRANEAN GULL *Larus melanocephalus*

P:10+1 m. S:-,E R:12

Ad postN complete June-Sept./Oct. **Ad preN** partial (head, B) Feb.-April. **PJ** (B) from shortly after fledging-late Sept; 1st preN (head,B) Feb.-April, moults to 2W June-Oct., in Wales well advanced by mid July. **(126,152,200,264,265,710).**

LITTLE GULL **L. minutus*

P:10+1 m. S:18-19,E R:12

Ad postN complete, timing variable, June-Nov.; some start P moult July **(581),** but in USSR not until Sept. **(126);** S1 falls when P6 growing, inner S descendant; R apparently start when P7/P8 dropped; non-breeding Ads in UK probably moult earlier (June). **Ad preN** partial (head,B) Jan.-May. **PJ** partial (B) from fledging-Nov. 1st preN partial (head, B, centre R, inner WC) Feb.-May, complete moult from June, probably finishing Sept., earlier than Ads. **(32,126,152,200,581,710).**

BLACK-HEADED GULL ***L. ridibundus*

P:10+1 m. S:20-21,E R:12

Ad postN complete June/July (Aug.)-Sept./Oct. (Nov.), a few start P moult during incubation, P duration c3 months.

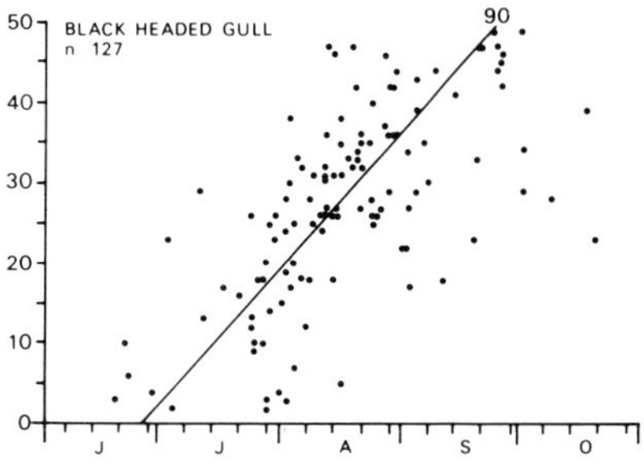

Often retain some NP on head throughout winter. **Ad preN** partial (head,B) Jan.-April. **PJ** partial (B), starts at fledging, usually finishing by Sept. 1st preN partial (head, B, inner WC) Feb.-April. 1st postN complete, starts c3 weeks ahead of Ad, P duration c2½ months. **(27,52,126,152,200, 437,476,485,670,710).**

COMMON GULL ***L. canus*

P:10+1 m. S:20,E R:12

Ad postN complete June-Oct., some start P moult during incubation, P duration c4½ months. **Ad PreN** partial (head, B) Feb.-May. **PJ** partial (B) from fledging-late Sept. 1st preN (B) Feb.-April, complete moult June-Oct. starting P at least 20-25 days before Ad. **(126,152,200,485,670,710).**

LESSER BLACK-BACKED GULL
*** *L. fuscus*

P:10+1 m. S:22,E R:12

Ad postN complete, timing variable, onset of P moult may be getting earlier (see also Herring Gull) and more birds undergoing complete moult in UK due to changing migration patterns. Some breeding birds start late May in Ross-shire (17% of those culled on eggs) and Lancashire during early incubation, but in Lancashire some still not started 5 Aug. Lancashire birds may suspend (16 of 130 June, 17/55 July) after early P moult start (I. M. Spence, pers. comm.). In Wales P started 2nd week July, when young well grown, in 1960's; recently shown that some Ads now start 1st week June, during later stages of incubation. In Iceland P start late May during egg laying, Pscore 25 by Sept. when birds migrate. In Scandinavia P start mid June *(intermedius)* or mid July *(fuscus)*, North Norwegian birds may not start before migrating. Some (most/all?) migrants suspend P moult, Ads in suspended moult in Surrey in early Sept. 10 of 13 in Surrey on 11 Nov. still moulting P and both on 20 Nov. (Pscore 48, 49). **Ad preN** partial (B) Jan.-April. **PJ** partial (B) finished Nov., partial moult (B) Jan.-April, complete 1S moult starts up to 2 months before Ad. **(50, 150, 200, 225, 270, 281, 390, 578, 581, 648, 710).**

HERRING GULL *** *L. argentatus*

P:10+1 m. S:21,E R:12

Ad postN complete, timing variable, onset of P moult may be becoming earlier in UK (also see Lesser Black-backed Gull). P start mid/late May Lancashire with slight overlap with egg laying, whole population starting within period of 50 days, P duration c4 months. In Wales started late June/early July in 1960s, recently shown most Ads now start early June. At Bristol Channel colonies 13 of 111 in period 30 April-9 May had started P moult. Some Ads suspend P moult during later stages of incubation/ when feeding young (Lancashire (I. M. Spence, pers. comm.) and 17 of 31 at Tenby, Pembrokeshire, early July). In Kent Ads recently finished P moult 21 Aug. Scandinavian birds start P early May at end of laying, birds in North a little later than those in South, P duration c6 months **(50)**, c3 months **(281)**; on White Sea, USSR P duration c4½-5 months; some presumed Scandinavian birds wintering in S England in P moult until Dec. Netherlands P duration c6 months. **Ad preN** partial (B) Jan.-March. **PJ** (B) July-Nov. Imm complete moult from April, P duration c4½ months in Netherlands. Some subAds breed whilst moulting P. **(50,64,152,200,204,225, 281,379,380,390,477,485,557,578,581,648, 670,710).**

GLAUCOUS GULL * *L. hyperboreus*

P10+1 m. S:24 (23,25),E R:12

Ad postN complete. Iceland starts end April (laying), finishes Nov./Dec., P duration 205 days, Alaskan birds moult more quickly; S start with S24 (descendant) when Pscore c15, S1 (ascendant) when Pscore 25-30, the 2 series meeting at cS16; R start when Pscore 28-30 (when most young have fledged), rapid, finishing at same time as P. **Ad preN** partial (B) Jan.-March. **PJ** partial (some B) summer-Oct. Imms start complete moult earlier than Ads; imms in UK Pscore: 20 Aug 36, 17 Sept. 44). **(126,152, 200,270,285,389,710).** ICELAND GULL *L. glaucoides* moult largely unstudied but probably similar to Glaucous Gull. 4 Ads Greenland Pscores 18 June 14, 15; 25 June 14; 26 July 29. I Ad Iceland 28 July Pscore 47.

GREAT BLACK-BACKED GULL
** *L. marinus*

P:10+1 m. S:(23-25),E R:12

Ad postN complete. Wales start June-Aug. when young well grown, finish mid Sept.-Dec. In Sussex Ads nearing end of P moult in mid Dec. Iceland start late May, finish early Dec., P duration c188 days; S24 shed when Pscore 8-10, S1 when Pscore 25-30, descendant and ascendant meeting at cS16; R starts when Pscore 43-45, basically centrifugal. Est P duration 7 months Netherlands. **Ad preN** (B) Jan.-April. **PJ** (B) ends Nov., Imms start complete moult up to 2 months ahead of Ad. **(152,200,225, 270,485,670,710).**

KITTIWAKE ****Rissa tridactyla**

P:10+1 m. S:-,E R:12

Ad postN complete May/June-Sept./Oct.; generally regarded that P start after young fledged, but recent UK data indicate that some birds start moult during the early stages of incubation and then suspend; of 3 birds caught on eggs in Wales on 10 June 1978 2 had dropped P1; of 31 caught on the Isle of May, Scotland 11-16 July 1982, 16 were in suspended moult (14 had replaced inner 3P, 2 had replaced 2P), all of 12 caught 17-22 July were in active moult (having resumed after suspension?), mean moult scores 11-16 July: 14.4, 17-22 July: 17.3 (H. Galbraith pers. comm.); in USSR c½ P replaced by early Aug. but marked individual variation; non-breeding Ad in Lancashire P duration c125 days. **Ad preN** (B) Feb.-May. Complete Imm moult before Ad, in Lancashire P duration 106 days. **(38,126,152,200,285,485,710).**

STERNIDAE

The moults of this family have been rather inadequately studied (Stresemann 1963). Primary moult patterns are complicated, frequently having two (occasionally three) active moult centres in each wing. Moult is usually arrested before breeding, the next post-nuptial moult starting at P1. The tropical Fairy Tern *Gygis alba* is unusual in that it suspends during breeding, thus having truely 'serially descendant' moult (Ashmole 1968. Dorward 1963). A number of 'species' described by early workers (e.g. *Sterna portlandica, S. pikei, S. havelli)* are now known to be non-breeding/immature plumage of other species, the *'portlandica'* plumage apparently being the normal dress of immature birds of several *Sterna* and at least one *Chlidonias* species, being the result of incomplete moult and bleaching/wear of old feathers (Dwight 1901, Grant *et al.* 1971, Haverschmidt 1972). Additional work is required to further elucidate the timing and extent of moult in both Adults and Immatures.

SANDWICH TERN **Sterna sandvicensis*

P:10+1 m. S:-,E R:12

Ad postN complete; B starts during incubation, P start mid July-late Aug.; UK Pscores: 16 Aug. 14,18,19,22,24, 19 Aug.17, 11 Sept. 23,24; may suspend during migration; P moult complex in African winter quarters with 2 (occasionally 3) series concurrent; in Mauritania 2nd series starts at P1 when 1st series has passed mid-point, 2nd series slower, only 1 P growing at a time in each series, 3rd series may start at P1 when P1-P4/P5 have been replaced twice, some arrest/ suspend. **Ad preN** Feb.-April, possibly not distinct from postN? **PJ** complete, B, WC, some T and R from early Aug.; P from Dec./Jan. finish May-July, 2nd series starting June/July. **(111,158,201, 229,238,286,710).**

ROSEATE TERN **S. dougalli*

P:10+1 m. S:-,E R:12

Little studied. **Ad postN** complete, Pstart while feeding young June-July; 2, occasionally 3, series in winter quarters. **Ad preN** partial from Feb., finished on breeding grounds. **PJ** (B) autumn, complete moult summer of 2nd year, one bird Pscore 50 but R growing June, some retain some Juv P and other feathers. **(111,201,527,710).**

COMMON TERN ***S. hirundo*

P:10+1 m. S:19-20,E R:12

Ad postN complete usually start after young fledge, R start before P and many probably finish R before migration; most start P early July-late Aug. but variable; 29 birds in UK latter ½ Aug.-1st ½ Sept. av. Pscore 12.2, range 7-22, some, still not started in Aug. (Netherlands) and Sept. (UK, Morocco), suspend and resume in winter quarters; in Mauritania 43 of 332 Ads caught Sept.-Nov. showed 2 series P moult, 2nd series starting at P1 when 1st series had reached P6/P7; a few start 3rd series; Ads in South Africa April/May all new P; est P duration 1st series c200 days **(36),** total moult duration 6-7 months **(581);** S moult in 2 groups, S14-S19/S20 moult twice (at beginning and end of P moult) ascendant and descendant from centre of group, S1-S13 moult once ascendantly. **Ad preN** partial (B,R,WC) Feb.-March, not separable from postN? **PJ** partial (B) Aug.-Nov., complete moult Feb.-June in South Africa, extent variable. **(1,36,201,229,238, 429,463,498,527,581,671,710).**

ARCTIC TERN **S. paradisaea*

P:10+1 m. S:18,E R:12

Ad postN complete in winter quarters; P start late Sept.-early Nov. finish Feb.-March, only moulted once, P duration c60 days **(508)** 3-3½ months **(581); Ad preN** partial (B,R) late Feb.-March. Birds in active P moult in Aug. but no note of collecting locality or age **(126). PJ** 1W birds moult P from Dec.-Jan. until April/May, P duration longer than Ads, extent variable. **(111,113,126,201,429,508,581,710).**

LITTLE TERN **S. albifrons*

P:10+1 m. S:16,E R:12

Ad postN complete; breeding birds start P late June-Aug. when feeding young, suspend during migration and resume in winter quarters, 2nd series starts at P1 when 1st series has reached cP7, 3rd series may start in some birds when 1st series reaches P10; est P duration c150 days in Morocco (extrapolation from mean Pscores of 17.6 in mid Aug. and 26.8 in mid Sept., 1st series only). In Germany estimated that each P takes c30 days to reach full length. **PJ** partial (B, inner S, WC?) July/Aug.-Oct.; complete moult in winter quarters, P start late Sept.-early Dec. (later in North American birds?). **(1,21,111,229,238,342, 386,397,463,567,581).**

BLACK TERN **Chlidonias niger*

P:10+1 m. S:15,E R:12

Ad postN complete, B starts late May/June (early July in USSR), some start P July and suspend, other start later (4 of 12 in Morocco in Sept. had not started); sometimes 2 series P moult. **PJ** complete in winter quarters, starting Oct.-Nov., P start Dec.-Jan. **(1,111,126,201,229,463,581,629, 644,710).**

ALCIDAE

Moult little studied due to shortage of specimens, especially of known age. Studies of captive birds (e.g. Birkhead and Taylor 1977, Swennen 1977) may not be comparable to studies in the wild.

GUILLEMOT **Uria aalge*

P:10+1 m. S:20,E R:12

Ad postN complete, B starts before young leave nesting ledges; P dropped almost simultaneously late July/early Aug., flightless Ads recorded attending young in Wales; captive birds renewed P in av 63 days (42-90 days), probably cannot fly until P 70-80% full grown, thus flightless 45-50 days, outer P start growth a little after inners; S dropped simultaneously when P c½ grown, replaced very rapidly av 25 days (only 14 days in 2 captive birds); R dropped simultaneously when P c½ grown, replaced in c60 days; moult in all tracts finished Oct./Nov., late starting birds moult P and dorsal B more quickly than early birds. **Ad preN** (head) mid Jan.-mid March (captive), in wild some may start almost as soon as postN finished. **PJ** partial (B,WC) Aug./Sept.-Oct. while P,S, and R still growing. 2Y and 3Y birds may undergo complete postN moult earlier than breeding Ads. **(68,506,581,613,655,656,710).**

RAZORBILL **Alca torda*

P:10+1 m. S:21,E R:12

Ad postN complete, B starts late June/early July, P simultaneous Aug.-Oct.; S and R simultaneous, start after P, progress of moult more rapid than in Guillemot (in captive birds). **Ad preN** (B) March/April, duration 2-3/4 weeks. **PJ** (B,WC) July-Oct. while P,S, and R growing. 1st preN March/April-May; 1st postN June-Aug./Sept. **(317,506,581,613,655,710).**

BLACK GUILLEMOT **Cepphus grylle*

P:10+1 m. S:17-19,E R:12 (13,14)

Ad postN complete; B starts July, P and S simultaneous from Aug., R simultaneous later, finish Oct./Nov. **Ad preN** (B) Dec./Jan.-March/April, duration c2 months. **PJ** (B) Aug./Sept.-Oct./Nov. (Dec.); 1st preN March/April-May/June. **(13,505,506,710).**

PUFFIN *Fratercula arctica*

P:10+1 m. S:-,E R:16

Ad postN partial (B), Aug.-Sept. (some possibly have complete moult). **Ad preN** complete in late winter but timing variable from Oct.-April; P simultaneous but a very few birds retain some outer P; some return to breeding colonies with outer P still growing April/May; all S and some R dropped when P just breaking throught sheath, remaining R later. Colourful bill plates and eye ornaments shed immediately after breeding and regrown in spring. **Juv** plumage retained until spring/summer of 2nd year, some start P March or earlier, others still in old plumage July/Nov. (some captive birds moulted P twice in 2nd and 3rd years and once in 4th year); some moult dusky face in spring, others retain it. **(226,227,506,581,613,710).**

COLUMBIDAE

Moults complicated due to differences in starting times and varying frequency of birds suspending, even within single populations. In some studies the age of some birds has almost certainly been incorrectly determined (see Insley *et al.* 1980). Further studies may well reveal that current descriptions for some species are incomplete or misinterpreted.

ROCK DOVE *Columba livia*

P:10 S:12 (13 r.),D R:12

Ad complete; in UK June-Nov?, Pscores 4 Sept. 31, 6 Sept. 32, 10 Oct. 42; Feral birds moult throughout year (most April-Dec. in Manchester); in USSR Ad start April/May, finish Oct./Nov., duration c5½-6 months, overlaps breeding season, heaviest intensity of moult Aug./Sept. after breeding; inner S (S10-S12) dropped first rapidly, other S ascendant and descendant from S1 and S9; R basically centrifugal, but some irregular. **PJ** complete, early fledged young finished Oct/Nov., others suspend/arrest during winter, in spring of 2nd year may continue from where suspended or start anew at P1, some birds do both and so 2 active series of moult at one time (in USSR). **(315,392,710).**

WOODPIGEON ****C. palumbus*

P:10 S:12 (13 r.),D R:12

Ad complete; most start P April/May, P usually dropped one at a time; 8-12% suspend whilst breeding June-Aug., resume Sept.; earliest birds finish Oct., most suspend again during winter so only 5% active P moult in Feb., resume in spring; min duration P moult 8 months (some probably take 10 months). A few Ad may retain 1-2 outer P until next moult. **PJ** complete, starts when c6 weeks old; 70% moult 2 or even 3 P at once in Sept, but only 15% do so in early Nov., when first birds start to suspend; 5% still in active P moult Jan./Feb.; majority, which suspend, resume March/April, usually before Ad; some (late fledged?) may not start P moult until spring. Some 1S birds start moult at P1 at about same time as Ad but before they have finished PJ moult. PJ R sequence 2-1-6-3-4-5; S moult centripetally from S1 and S12 but may be incomplete/irregular. Variations in published accounts of timing, duration and extent of moults may be partly related to differences in food supply between different study areas. **(40,72,108, 343,393,412).**

STOCK DOVE ***C. oenas*

P:10 S:-,D R:12

Ad complete (May) June/July-Nov. (Dec.) overlapping with breeding; in USSR start May, most finish Oct. **PJ** complete from May/June, some not finishing until 2nd year. R replaced rapidly while P1-P3 moulting; most B replaced during later stages of P moult. Suspended P moult not recorded in USSR but 4 cases in UK: 1Y 3 Oct. Pscore 30, 1Y 15 Oct. Pscore 40, 2Y 5 May Pscore 35, not aged 27 Jan. Pscore 45, and 3 birds with Pscores of 37,45 and 48 in late May/mid June probably continuing after an earlier suspension. **(126,196,710).**

COLLARED DOVE

*****Streptopelia decaocto*

P:10 S:11 (12 r.),D R:12

Ad complete; March/June-Aug./Oct. (Dec.), P duration 75 days Hampshire but in Cheshire/Manchester c10 months; S start when P5 dropped. **PJ** complete, av start 1st week July but considerable scatter

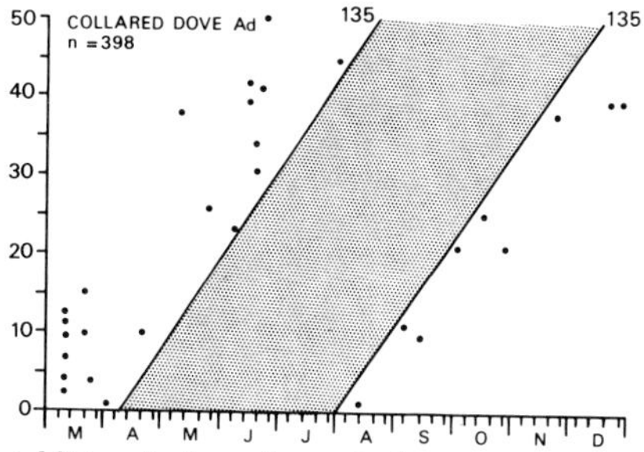

Additional data from Insley, Young and Dudley (1980)

of starting dates; in Hampshire mean increase of Pscore 0.39/day (50 retraps), est P duration 128 days; some suspend P during winter, of Juvs showing P moult in Feb., 50% suspended in UK. In Netherlands 50% of birds (age?) collected in Jan. were moulting wing and R. In captive Ad P duration 4 months, S duration 3 months; PJ start when 4-5 weeks old. **(110,125,271,289, 322).**

TURTLE DOVE *******S. turtur*

P:10 S:11 (12 reduced),D R:12

Ad postN complete, starting towards end of breeding reason, B starts June/July; most apparently start P late July/Aug. (P1 up to P4 new), then suspend before migrating; finish in winter quarters. **PJ** complete; B and sometimes P start July/Aug. before migration in 1st broods, 2nd broods do not start until reaching winter quarters; moult finished winter quarters, by March/April in West Africa. 8 of 10 in Iberia and 10 of 16 in Crete in suspended P moult (age not stated). **(42,329,359,383,386,391,612,581).**

CUCULIDAE

CUCKOO *******Cuculus canorus*

P:10 S:9,E R:10

Moult complex. **Ad postN**, B partly moulted before migration June/July and occasionally some wing and R; complete moult in winter quarters; P in two series, P1-P4 descendant, P5-P10 ascendant and alternate (viz. 9-7-5-10-8-6), the relative timing of the 2 series may vary; S moult centripetally from S1 and S9 with a tendency (especially in 1W) to stop before all replaced, thus some old centre S may be retained. Some Ad in B moult April/May. **PJ** (B) starts soon after fledgling; some start wing and R before migration, most undergo complete moult in winter quarters. Some in spring may retain in few old P. **(42,99,529,589,649,710).**

STRIGIFORMES

Most detailed studies have been of captive birds and the duration of moult may not always be applicable to those in the wild. Further studies of wild individuals are required. Irregularities in the moult have been recorded in individuals of several species which have exhibited obvious signs of disease and/or starvation (Ginn and Glue 1974, Hardy *et al.* 1981). For discussions on tail moult see Mayr and Mayr (1954), Steinbacher (1955) and Verheyen (1956).

TYTONIDAE

BARN OWL **Tyto alba*

P:10 S:12 (+ 4 small),D R:12

Ad in B moult throughout year; P moult usually starts at P6 and thereafter ascendant and descendant; captive birds may take more than 1 year to complete with 2 series active, thus P6,P7,P8 moult before P1,P2,P3 of previous series; insufficient data from wild birds to determine P duration but in UK active P moult recorded in 10 birds July-Dec.; S in 3 groups S1-S4 and S5-S7 ascendant, S12-S8 descendant; R irregular and slow. **PJ** B moult begins whilst still in nest, flight feathers start in 2nd summer, perhaps slightly earlier than Ad. **(43,224,457,462,519,581,606,710).**

STRIGIDAE

SNOWY OWL **Nyctea scandiaca*

P:10 S:19,D R:12

Ad postN complete June-Nov.; P usually start during incubation with P6, thereafter ascendant and descendant, duration c3 months; S in 3 groups? **PJ** partial (B, some WC), starts before P,S,R, full grown, finishes Sept./Oct. **(126,513,581,677,710).**

LITTLE OWL *******Athene noctua*

P:10 S:14,D R:12

Ad postN complete, ♂ apparently starts before ♀, (June) July-Nov.; P descendant

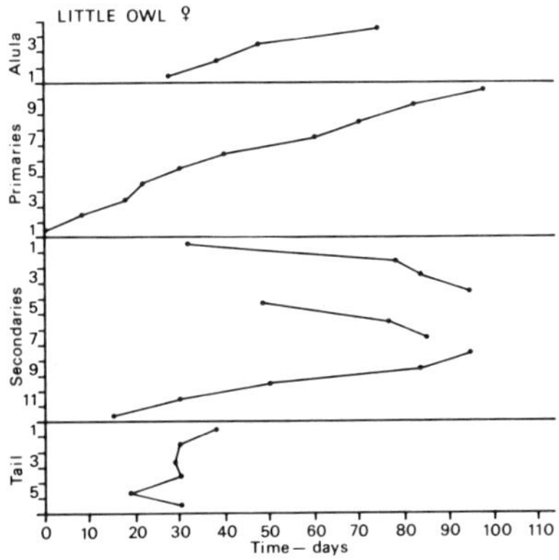

Sequence and timing of moult in a captive female Little Owl. Redrawn from Piechoki (1968b)

from P1; S in 3 groups S1-S4 and S5-S7 ascendant, S12-S8 descendant; R sequence variable and more or less simultaneous dropping when P4/P5 growing. **PJ** partial (B, some WC) Aug.-Nov., flight feathers in 2nd summer. UK Pscores: 18/20/25 July 2,2,3,5,8,17; 8/11 Aug. 14,21,29; 25 Aug. 26; 2/3/9 Sept. 29,29,42; 17 Sept. 33; 9 Oct. 47. **(42,231,232,337,355,461,581,639,710).**

TAWNY OWL **Strix aluco*

P:10 S:14,D R:12

Ad postN complete; P descendant from P1, starting early June UK (May-June central Europe), failed breeders may start earlier than successful breeders, av P duration 77 days (UK); S moult in 3 groups S1-S4 and S5-S7 ascendant, S12-S8 descendant but S11 may drop before S12; R start in later stages of P moult centripetal/irregular and nearly simultaneous (1♂ dropped all R in 8 days, 1 ♀ in 4 days — captive birds); B continues until Dec. (a few -Jan./Feb.). **PJ** (B, some WC) may start before P,S,R full grown, finishes Oct.-Dec. **(191,224,457,581, 617,710).**

LONG-EARED OWL *Asio otus*

P:10 S:12,D R:12

Ad postN complete June/July-Dec., ♂ perhaps earlier than ♀; P descendant from P1, P duration 75-97 days; S in 3 groups S1-S4 and S5-S7 ascendant, S12-S8 descendant, S1 falls c20 days after P1, S6 c65 days after P1, S finish about same time as P; R more or less simultaneous 40-50 days after P (when P6/P7 moulted); Alula descendant, A1 dropped 3 days after P1, A4 77-103 days after P1 (all from captive birds). **PJ** (B,WC) starts when wing and R are starting to grow, often not finished until Nov. UK Pscores: 3/5 July 3,3; 22/27 July 5,10; 15Aug. 10; 6 Sept. 41. **(54,126,460,581,710).**

SHORT-EARED OWL *A. flammeus*

P:10 S:-,D R:12

Ad postN complete, June/July-Nov. in UK; P descendant from P1, P duration 58-89 days; R dropped more or less simultaneously when P6-P8 dropped; B starts after P but finishes about same time, duration 74 days (captive). **PJ** (B,WC) starts when wing and R starting to grow, often not finished until Oct. UK Pscores: 29 June 12, 22 July 5, 13 Aug. 17, 5 Sept. 30. **(156,561,581,710).**

CAPRIMULGIDAE

NIGHTJAR *Caprimulgus europaeus*

P:10 S:13,D R:10

Ad postN complete, B starts July and inner S may be moulted before migration, finished in winter quarters late Oct.-March; S ascendant and descendant, may stop before centre S replaced; R sequence 1-2-3-5-4. **PJ** (B) Sept./Oct., complete moult in winter quarters, probably starting a little later than Ad. **(42,126,561,581,650,710).**

APODIDAE

SWIFT ***Apus apus*

P:10 S:8 (9,10 r.),E R:10

Ad postN complete; B starts Aug. (July) before migration and at least some start P (20% moulting P1 (a few P1 and P2) Dorset); finish moult in winter quarters; in Central Africa P start mid Aug., continue to late Feb.-mid March, P duration 6-7 months; 20-30% do not renew P10 (a few may arrest earlier, 1 in UK with P8-P10 old May), which is then retained until following Sept./Oct., being dropped when next series reaches P3-P7; individual birds apparently retain P10 every 5th year on average. **PJ** (B only?), wing not moulted until 2nd autumn. **(128,451,581,683,710).**

ALCEDINIDAE

KINGFISHER **Alcedo atthis*

P:10 S:12 (13,14 r.),D R:12 (13,15)

Ad postN complete July-Nov./Dec.; P in 2 groups P1-P6, P7-P10 both descendant (1 confirmed case 2nd group starting with P6), moult concurrently but either group may

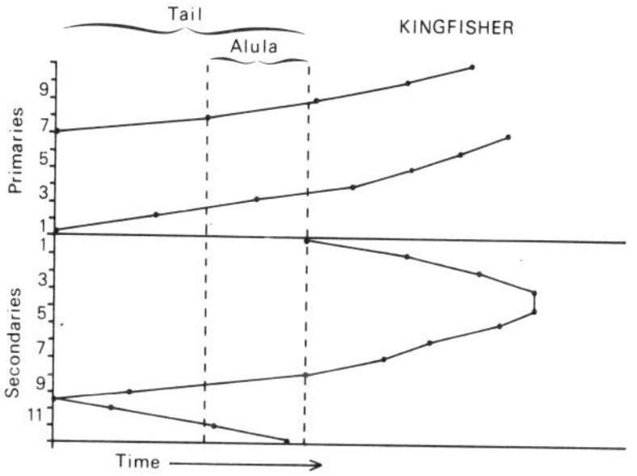

Sequence and timing of moult in the Kingfisher. Generalised diagram redrawn from Doucet (1971)

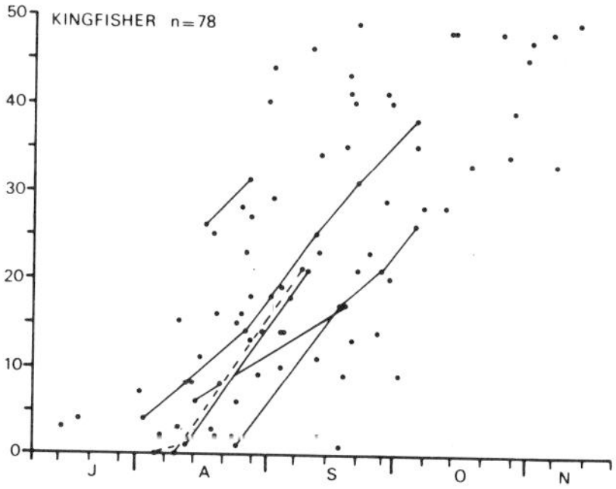

Dashed lines join date of last capture before start of moult with first plot of active moult for two retraps.

start first; estimated P duration 3¼ months France, UK retraps suggest c90-100 days; S moult from 2 centres at S10 and S1, the two series usually meeting at S4 (sometimes S3,S5); R irregular start at same time as P, more or less finished by time S1 drops. Late moulting birds may arrest/suspend during winter — the retained feathers apparently being the 1st to be replaced at next moult — needs confirmation. In UK birds in suspended/arrested moult: 25 Nov. Pscore 45, 28 Dec. Pscore 45, 7 April Pscore 20. **PJ** (B, most replace centre R but variable, many replace all?) July-Dec. **(142,341,569, 590,710).**

UPUPIDAE

HOOPOE **Upupa epops*

P:10 S:10 (11 r.) R:10

Little studied, published information somewhat conflicting. **Ad postN** complete in winter quarters, end in Jan./Feb. (March), but some start earlier B from late July/early Aug. (occasionally P1 and S1 also); S moult from 2 centres: ascendant from S1, ascendant and descendant from S7; some apparently suspend/arrest, thus 4 birds in South Spain/Canaries not moulting had P scores: 3 Sept. 20, 30 Jan. 35, 2 March 15, 25 April 15; 1 in Canaries 24 Feb. had retained old P3 and P4 (all other P new). **PJ** partial (B) usually some/all R, in winter quarters. **(126,196,243,568,571,581, 710).**

PICIDAE

The inner 2 (in some species 3) Primaries are much reduced in Juveniles and are moulted before, or soon after, leaving the nest. P10 is much reduced in Adults, longer in Juveniles (½ length of the longest Primary). 12 Rectrices but the outermost pair short and soft, lying above the 5th pair in woodpeckers, below the 5th pair (as in other birds) in Wryneck *Jynx torquilla*. Post-juvenile moult 'complete' except for Secondaries which are usually retained until 2nd year. (Chapin 1921, Heinroth 1916, Kipp 1956, Sibley 1957, Stresemann and Stresemann 1966, Sutter 1974, Verheyen 1957).

WRYNECK **Jynx torquilla*

P:9+1 S:10,E R:12

P10 r.(=½ longest) in Juv. m. in Ad.

Ad postN complete, rapid, July-Sept./Oct.; R centripetal; S from 2 centres, ascendant from S1, ascendant and descendant from S8; some birds start migration before moult finished: single birds at Gibraltar and Cyprus late Oct. with new P and active S moult, 1 Lincolnshire 25 Aug. with new P,S and all R except centre pr at stage 4. **Ad**

preN partial (B), some birds also replace some R and S in winter quarters before spring migration. **PJ** 'complete' except for S, starts before fledging, P1 when 22-24 days old, B starts when P5 dropped at 40-45 days old, finish July/Aug. **(43,126,196,581, 601,607).**

GREEN WOODPECKER

**** Picus viridis*

P:10 o.r. S:11-12,E R:12

Ad postN complete June-Nov.; S from 2 centres, ascendant from S1 (dropped soon

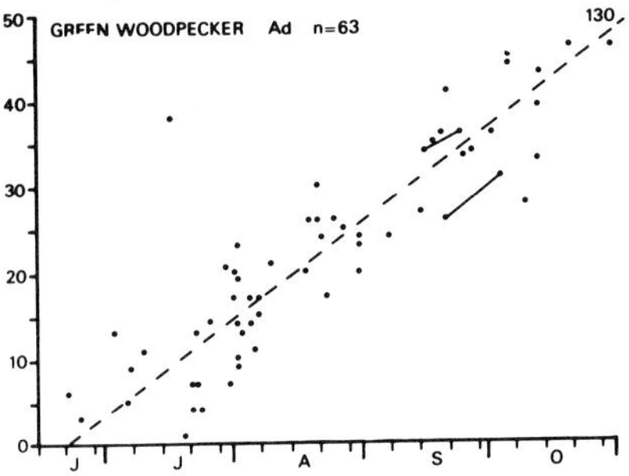

after P5), ascendant and descendant from S8 (dropped with P3); R1 last to be shed, otherwise centrifugal (but R6 may be dropped before R4 and R5). **PJ** 'complete' July-Oct.; S retained (sometimes 1 T replaced); P1 often dropped before

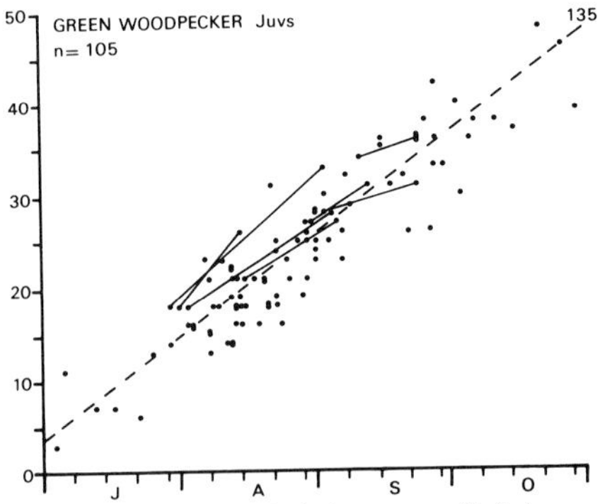

fledging, P2 drops 1-4 days post fledging. **(43,98,196,543,607).**

GREAT SPOTTED WOODPECKER

**** Dendrocopus major*

P:10 o.m.r. S:11-12,E R:12

Ad postN complete June-Nov. S from 2 centres, ascendant from S1,

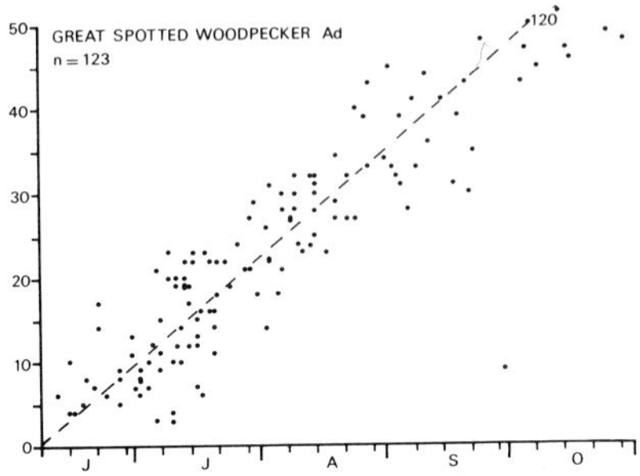

Date and Age/sex	Primary Coverts 1	2	3	4	5	6	7	8	9	10
23 Jan —	5	0	0	0	0	0	5	5	5	5
22 Feb 4♂	5	5	0	0	0	5	5	5	5	5
28 Feb 4♀	0	0	0	5	5	5	5	5	5	5
28 Feb 4♂	5	5	5	0	5	5	5	5	5	5
20 Nov 4♂	5	5	0	0	0	0	5	5	5	5
12 Dec -♀	5	5	5	0	0	0	5	5	5	5

Incomplete moult of Primary Coverts in Great Spotted Woodpecker Dendrocopus major *in U.K. (all birds had 10 new P.)*

ascendant and descendant from S8; R1 last to be shed, otherwise centrifugal? This species is unusual in that the Primary Coverts are not always shed in sequence with their respective Primaries and some may be retained through a moult (see table). **PJ** 'complete' June-Nov.; S retained; timing

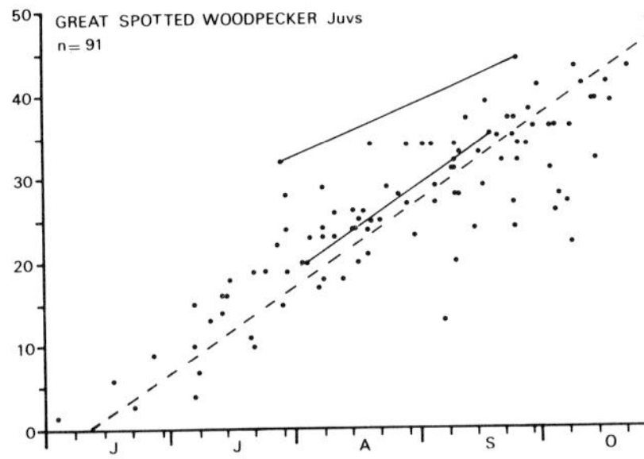

of P moult: P1 dropped at age 20 days (leave nest at c21 days), P2 at 22 days, P3 at 48 days, P4 at 62 days, P5 at 71 days, P6 at 89 days, P7 at 106 days **(499);** R irregular? R5 dropped at 40 days old, R6 at 72 days, R2 at 81 days, R3 at 84 days, R4 at 96 days, R1 at 118 days **(499)**. During 'irruption' years in Finland intensity of PJ moult reduced, most suspend/arrest. **(41,98,499,581,607,660, 661).**

LESSER SPOTTED WOODPECKER

****D. minor*

P:10, o.r. S:11-12,E R:12

Ad postN complete June-Sept.(Nov.); S from 2 centres, ascendant from S1, ascendant and descendant from S8: R1 last

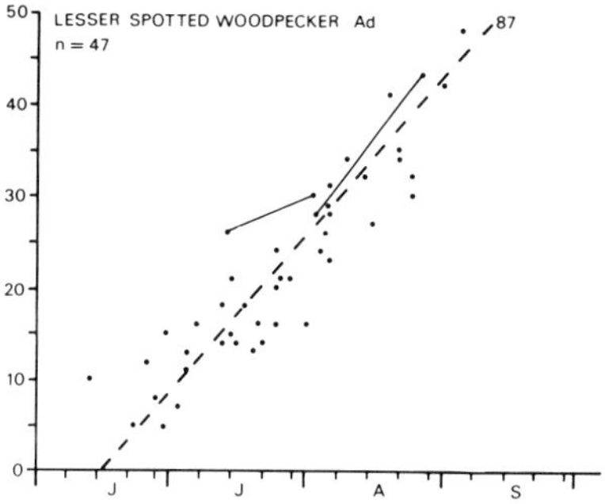

to be shed, otherwise centrifugal? **PJ** 'complete' June-Sept./Oct. S retained; P1 and P2 dropped before/shortly after fledging. **(41,581).**

PASSERINES PASSERIFORMES

The basic pattern of pterylosis described for *Passer* species (Clench 1970) fits all Passeriform families (Clench 1973).

ALAUDIDAE

WOOD LARK **Lullula arborea*

P:9+1 m. S:6,E T:4 R:12

Ad postN complete, start early/mid July; most moulting early Sept., finish late Oct. in USSR. **PJ** complete, birds in moult to late Oct. in USSR. P9 3-6mm shorter than longest PC in Ad, 1-5mm longer than longest PC in Juv. **(126,136,710).**

SKYLARK ****Alauda arvensis*

P:9+1 m. S:6,E T:4 R:12

Ad postN complete early July/early Aug.-mid/late Sept.; P duration c58 days in Lincolnshire. **PJ** complete early July/late Aug.-late Aug./mid Oct. **(117).** *See figure in next column.*

SHORE LARK **Eremophilia alpestris*

P:9+1 m. S:6,E T:4 R:12

Ad postN and **PJ** complete early autumn (late July-Sept). **(126,608).**

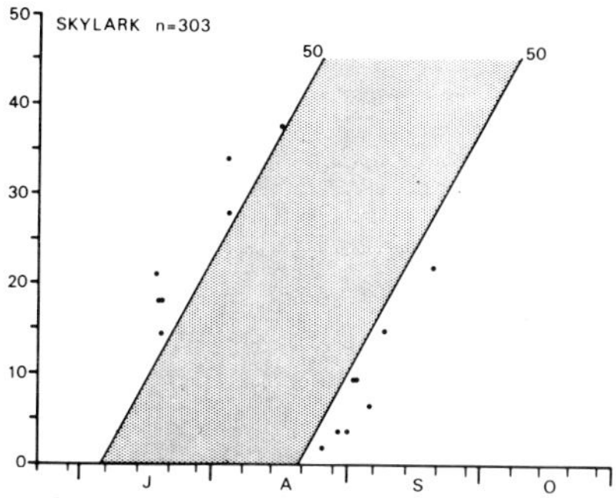

Includes data from Davies (1981)

HIRUNDINIDAE

SAND MARTIN **Riparia riparia*

P:9+1 v. S:6,E T:3 R:12

Ad postN complete; c2% UK ad start P moult (P1-P2/P3) before migration (20 July-17 Sept., most late Aug.-early Sept.), others replace some T and/or R but not P; P1 and P2 often moulted together; S1 dropped with P4; active moult may continue during the early stages of migration but probably suspended later, finishing in winter quarters. In Kenya and Uganda moult late Oct./Nov. — mid March/mid April. Est P moult duration in Zambia 135 days (D. M. Francis, pers.

Primary moult of Sand Martins caught in the Lusaka area of Zambia 1972-75 between mid-November and mid-February: data provided by D. M. Francis.

comm.). Of 445 Ad from West Tien Shan late Aug.-early Oct. 145 in suspended P moult, most having replaced P1-P2/P3, a few P1-P6. **PJ** complete in winter quarters; no records of birds starting in UK, a few records from Spain and Greece. P duration c141 days in West Africa, c120-150 days in East Africa. **(53,144,187,332,358,540).**

SWALLOW ****Hirundo rustica***

P:9+1 v. S:6,E T:3 R:12

Ad postN complete; a variable number start P before migrating, a greater proportion doing so in the south of the breeding range (eg. Switzerland 3%, West Germany 5-10%, South Spain 19%); one ♀ replacing P1 whilst rearing young in Iraq, one probable record in UK; suspend and finish in Africa, some start/resume moult in Africa before reaching winter quarters. 6.5% suspended after renewing P1-P3 in West Tien Shan. Timing and duration of moult in Africa

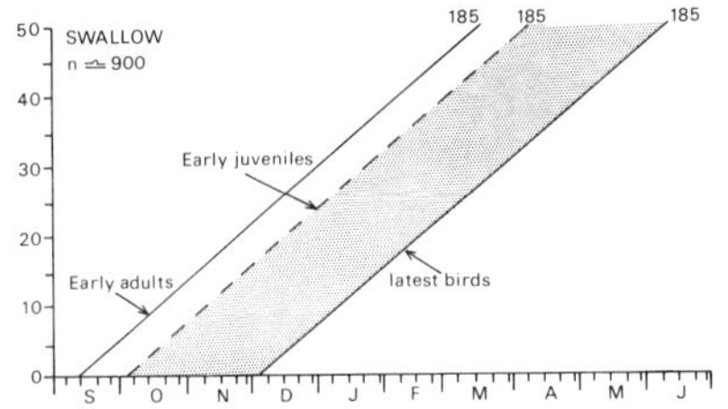

Primary moult of Swallows caught in the Lusaka area of Zambia 1972-75: data provided by D. M. Francis.

variable: at Lusaka, Zambia P duration c185 days, T start soon after P (av when Pscore 3.7) and usually moult in sequence 8-9-7; S start when Pscore 13-23 usually finishing before P, R start when Pscore 16-30 generally centrifugal but variable; elsewhere in Zambia T start after S. P duration c150 days in East Africa. **PJ** complete, usually in winter quarters, only a very few appear to start P before migration. In Central Africa Juvs moult 4-6 weeks later than Ad, and c3 weeks later in East Africa, but both at the same time in South Africa. By the time that birds leave South Africa in April a considerable number have not yet finished P and R moult. **(53,82,121,122,143, 179, 187, 248, 290, 294, 332, 357, 361, 367, 468, 469, 473, 488, 540, 583, 708).**

HOUSE MARTIN ***Delichon urbica***

P:9+1 v. S:6,E T:3 R:12

Ad postN complete; P may start before migration, in Switzerland 5% in active moult or suspended after replacing P1 or P1 and P2 in early Oct; suspended P moult also recorded in Iberia (3 of 5) and Germany, but not evident in 74 Ad from West Tien Shan; moult finished in winter quarters. Complete moult late Oct./Nov.-(Feb.) March in East Africa. **PJ** complete in winter quarters, fewer Juvs start P before migration, 0.1% in Switzerland, 2 of 7 in Portugal. **(187,298,332,357,359,540,708).**

MOTACILLIDAE

TREE PIPIT ****Anthus trivialis***

P:9+1 m. S:6,E T:3 R:12

Ad postN complete July-Sept. P duration c35 days. In Belgium ♂ start before ♀. **PreN**

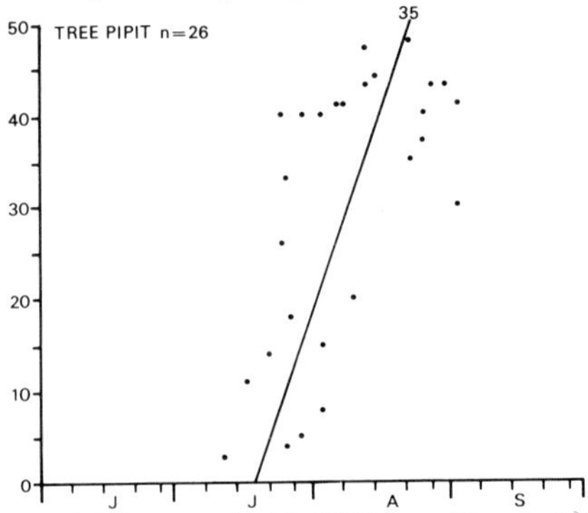

partial (B, some R, T, GC?) Jan.-March. **PJ** partial (B, some inner GC, some T, sometimes centre R) July-Sept. **(221,237, 298,331,641,710).**

MEADOW PIPIT *****A. pratensis***

P:9+1 m. S:6,E T:3 R:12

Ad postN complete, July-Sept./Oct. P duration 40-50 days in Finland, c50 days in

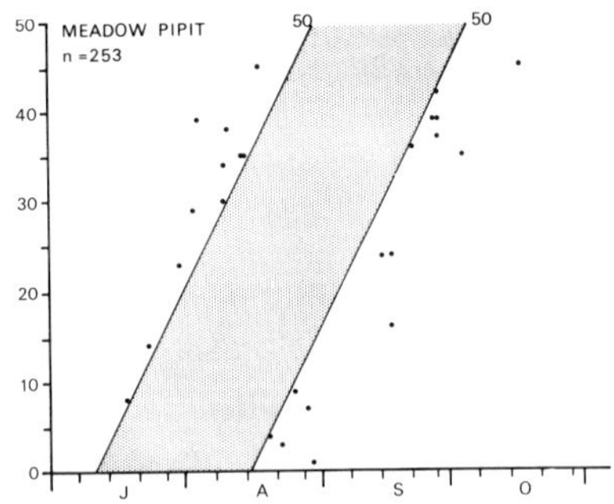

UK. **PreN** partial (B centre R, T, some GC) Jan.-March. **PJ** partial (B, some T and GC, sometimes centre R) July-Oct. **(221,234, 236,237,327,710).**

ROCK/WATER PIPIT ***_A. spinoletta_

P:9+1 m. S:6,E T:3 R:12

Ad postN complete, June/July-Sept. On Fair Isle _kleinschmidti_ starts B after P1 and P2 moulted (but in USA _alticola_ starts B before P), then T (8-9-7) and GC; R start when P3 replaced, centrifugal; S start when P c½ replaced (P6 in USA). In USA ♂ start before ♀; P duration c42 days in UK,

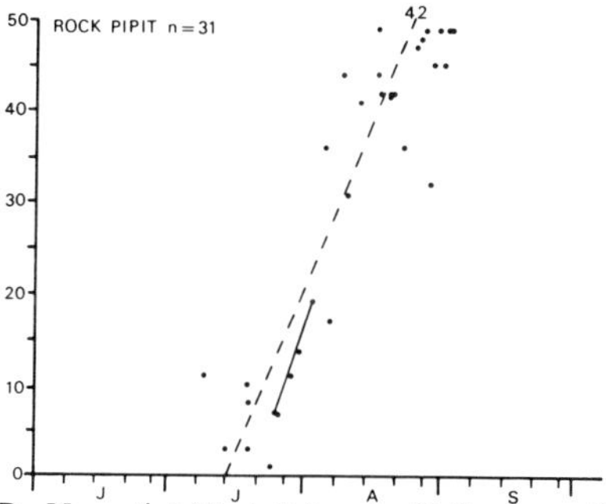

PreN partial (B, 1-2 (rarely 3) T, centre R, some WC) but the extent of moult varies greatly between populations, in UK populations absent or incomplete. **PJ** partial (B,T, some GC) Aug.-Sept. **(221,354,647,700,710).**

YELLOW WAGTAIL ***_Motacilla flava_

P:9+1 m. S:6,E T:3 R:12

Ad postN complete July/Aug.-Aug./Sept., in UK av starting date 19 July, av finish 31

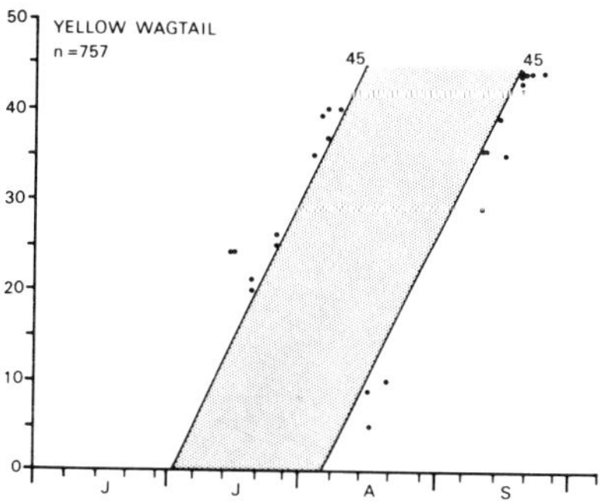

Aug.; all feather tracts moult simultaneously; P duration c43-45 days; R moult irregular, 37% centrifugal, 17% simultaneous, 33% R1 and R6 together, 13% anomolous and asymetrical; migration apparently starts before moult finished. **PreN** in winter quarters Jan./Feb.-

Moult of Greater Coverts by Juvenile Yellow Wagtails Motacilla flava _in UK_*

N =	0	1	2	3	4	5	6	7	8	9
% =	7	0	1	2	2	4	10	15	12	48

N = number of old GC retained
% = percentage of 582 Juvs examined
*after Hereward 1979

March/April, duration 12-13 weeks, partial (B, T, some WC, R) contour feathers start c2 weeks after and finish c3 weeks after T and R; finished before spring migration. **PJ** partial (B, occasionally some R and T, variable number WC); in UK over ½ moult at least 1 GC, a few moult all (see table). In Sweden less than 5% moult any GC. **(114,233,237,246,320,327,608,712).**

GREY WAGTAIL **_M. cinerea_

P:9+1 m. S:6,E T:3 R:12

Ad postN complete, July-Sept. P duration c65 days; R irregular. **PreN** partial (B, some

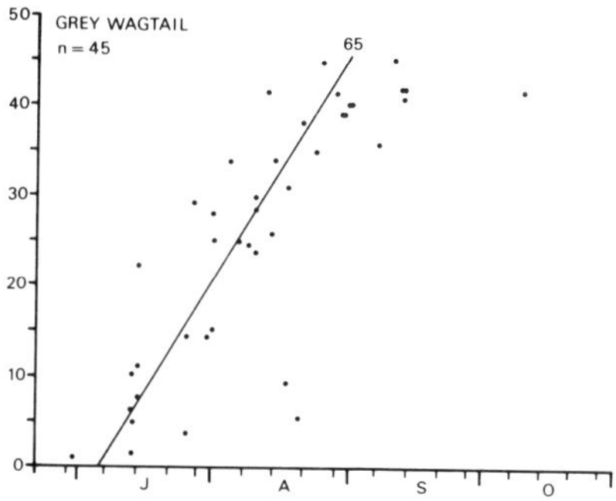

T and centre R, occasionally all R and some WC) Feb.-March. **PJ** partial (B, usually 3-4 GC (some occasionally replace all, late broods may not replace any), some replace some/all T (irregular), a few replace some/all Alula, sometimes centre R replaced July/Sept.; start soon after fledging (c2 weeks?). **(298,710).**

PIED/WHITE WAGTAIL ****_M. alba_

P:9+1 m. S:6,E T:3 R:12

Ad postN complete, early July/mid Aug.-late Aug./late Sept.; S1 shed with P5/P6, S finish 3-25 (av c14) days after P; T moulted between dropping of P2 and P8; in Somerset R moult lasted throughout total moult period, in Scotland start between shedding P4 and P5 finished before P8 full grown; GC replaced in 3 groups: 9-7, then 6-

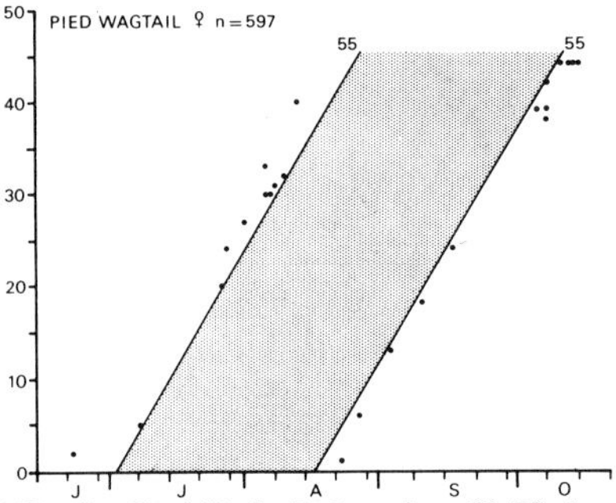

4(3), finally 2(3)-1; P duration 68-78 days Somerset, 73 days in Scotland, c66 days in Berkshire; no difference in rate of P moult between ♂ and ♀ in Somerset and lines fitted by eye to BTO data suggest P duration for ♂ and ♀ 55 days; in Finland migratory *alba* P duration 45-50 days, in Sweden some ♀ with late broods may start moult before young fledge. **PreN** partial (B, some T and WC, centre R) Jan.-March. **PJ** partial late June-late Aug. (B, some R), GC replaced in same order as Ad but distal 2-3 usually retained (late hatched birds may retain all, in Somerset ♀ retained more old GC than ♂); T usually replaced but in some (late hatched?) retained or partly replaced; av duration PJ moult 51 days (Somerset) but early starters take longer than later birds (60:42 days); in Scandinavia *alba* occasionally moult some T (very rarely all), usually 1 or a few (occasionally up to 7) inner GC replaced. **(24,26,183,233,234, 320,327,336,454,608,710).**

BOMBYCILLIDAE

WAXWING *Bombycilla garrulus

P:9+1 m. S:6,E T:4 R:12

Ad postN complete, Oct./Nov. in USSR. A few in Scandinavia suspend, retaining a few inner S, 1 Swedish bird active moult 10 Feb. Pscore 38, 1 Lappland 6 June Pscore 44. In UK: 22 Oct. Pscore 38 (P10 old), S1 and S2 new rest old; 3 Nov. P all new, S1-S3 new, S4 score 2, S5 score 1, S6 score 0. **PJ** partial (B,WC?) from late Aug., complete moult Aug.-Nov. of 2nd year. **(126,608).**

CINCLIDAE

DIPPER ****Cinclus cinclus*

P:10 o.r. S:6,E T:3 R:12

Ad postN complete late May/early June-late Aug./Sept.; P1-P5 dropped more or less simultaneously, when P1-P3 full grown (usually before P4,P5) P6 dropped, thereafter descendant each P dropped when preceding P c⅓ grown; T start with T8 when P5 (occasionally P6) dropped, sequence of T9 and T7 variable (in Europe in sequence 9-8-7); S1 dropped with P7, sequence of S2-S6 variable but none dropped until (or after) P9, S finish c15 days after P; R rapid, most drop all R simultaneously (38 of 52 in Scotland), others drop 2-4 prs simultaneously, R duration c50 days (Scotland); P duration av c70 days (Scotland) but considerable

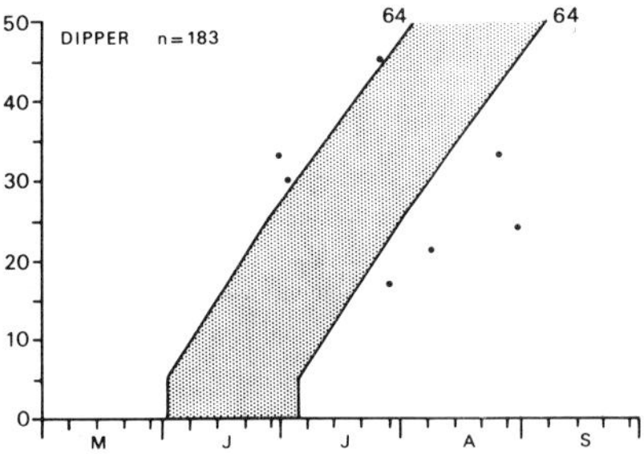

Additional data from Galbraith, Mitchell and Shaw (1981) are included.

variation thus P duration of 21 retraps 48-98 days; ♂ may start moult a little earlier than ♀. **PJ** partial (B, inner 2-4 GC) June-Sept., start when c60 days old. Although flightlessness reported in Czechoslovakia (and in *Cinclus mexicanus* in USA) no flightless birds found in Scotland, but secretive and reluctant to fly. **(45,184,491, 581,591,710).**

TROGLODYTIDAE

WREN ****Troglodytes troglodytes*

P:10 o.r. S:6,E T:3 R:12

Ad postN complete early July/mid Aug.-early Sept./mid Oct.; P duration c60 days. **PJ** partial, variable; those starting in July tend to renew B,R, some (usually 5) GC, and sometimes T but late fledging birds may only renew B; those renewing R and GC take c60 days; T and GC tend to moult simultaneously. **(239, 298,622).** *See figure on next page.*

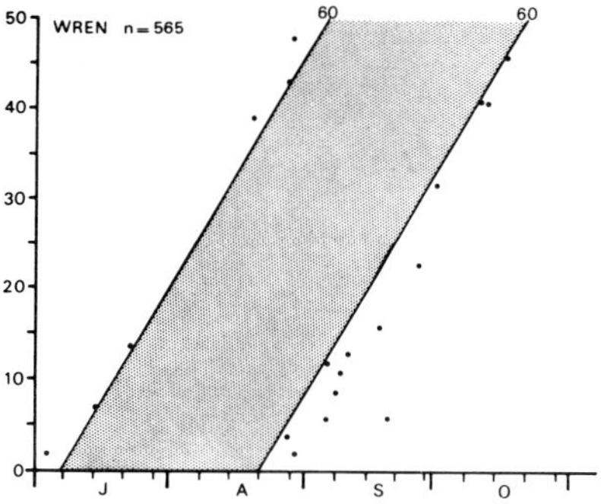

PRUNELLIDAE

DUNNOCK **** *Prunella modularis*

P:9+1 v.r. S:6,E T:3 R:12

Ad postN complete; most start P mid July/mid Aug., finish late Aug./mid Oct., earliest birds finish P up to 14 days before late birds start; moult starts after breeding, birds in South England, Scotland and North England consistently start and finish

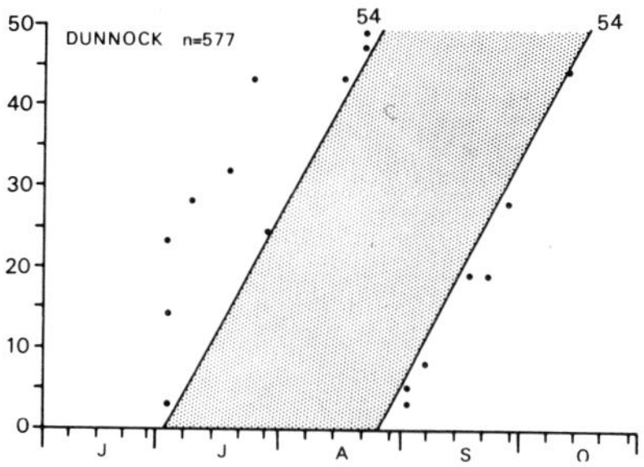

moult earlier than those in Central England and Wales; P duration 54-60 days (10P), 48-54 days if only 9P scored; S start when Pscore 18-30 (when P5 or P6 dropped) most finishing before P; T usually start when Pscore 10-25 (34% moult all 3 in quick succession, 27% drop T8 first, rest variable); R start at Pscore 1-25 (c60% centrifugal, c16% more or less simultaneous, rest irregular). **PJ** partial (B,T, some inner GC, sometime R) late June-mid Oct. **(73,190, 298,550,622,699).**

TURDIDAE

ROBIN **** *Erithacus rubecula*

P:10 o.m.r. S:6,E T:3 R:12

Ad postN complete, early June/early Aug.-

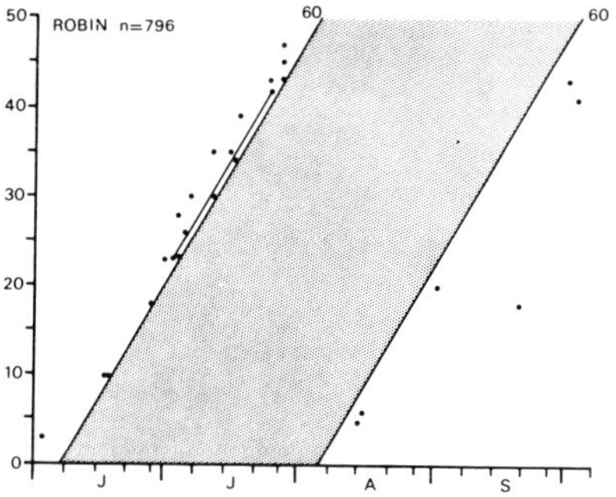

late July/late Sept.; P duration variable: in UK c60 days, in Germany c80 days; R centrifugal R1 dropped shortly after P1; S in sequence 8-1-7-2-9-3-4-5-6, S8 falls shortly after P4 and S6 full grown shortly after P9 (P10 finishes earlier); moult on breast starts after P3 dropped and before P4. **PJ** partial (B, some inner GC, occasionally some T) mid June-early Oct., starting when 6-7 weeks old, duration c55 days. **(347,494,544,550,622).**

NIGHTINGALE

** *Luscinia megarhynchos*

P:10, o.m.r. S:6,E T:3 R:12

Ad postN complete before migration late June-early Sept. P moult rapid (c45 days)

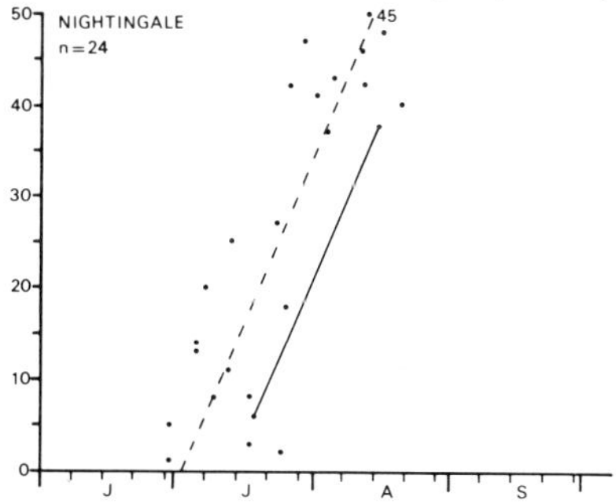

but not as quick as the Thrush Nightingale *Luscinia luscinia* which takes 30-35 days. **PJ** partial (B, some inner GC, occasionally some T) July-Sept. No **PreN** moult. **(59,544,710).**

BLUETHROAT * *Luscinia svecica*

P:10, o.m.r. S:6,E T:3 R:12

Ad postN complete after breeding and before migration, mid July/Aug.-Sept. for *svecica;* P duration 40-45 days in Finland. In Morocco one *cyanecula* just starting B 19 Sept., P,S and R very abraded. **PreN** partial (head, breast) Feb.-March/April. **PJ** partial (B, some inner GC) July-Sept. **(233,234,327,386,710).**

BLACK REDSTART *******Phoenicurus ochruros*

P:10, o.m.r. S:6,E T:3 R:12

Ad postN complete before migration, end July/Aug.-Sept./Oct.; P duration c50 days; peak of B moult late Aug. in Germany.

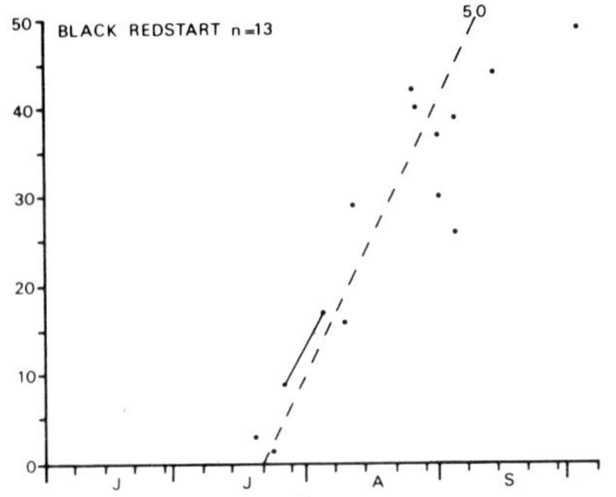

Details of **preN** moult (if any) unknown. **PJ** partial (B, variable number GC 0-4) July-Sept. **(298,544,710).**

REDSTART *******P. phoenicurus*

P:10, o.m.r. S:6,E T:3 R:12

Ad postN complete before migration June/July-early Aug./Sept.; P duration c40 days; peak of B moult late July in Germany.

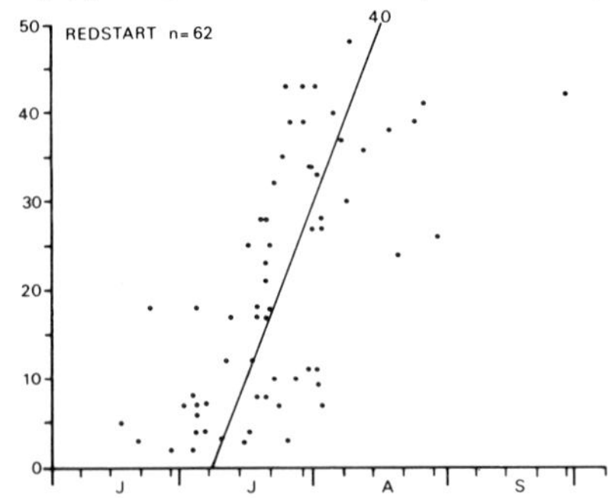

No **preN** moult, spring plumage acquired by abrasion. **PJ** partial (B, 1-2 (sometimes 0) GC, occasionally some R) end June-Sept. **(234,298,544,710).**

WHINCHAT *******Saxicola rubetra*

P:10, o.m.r. S:6,E T:3 R:12

Ad postN complete before migration, July/early Aug.-late Aug./Sept.; P duration c50 days in UK and Finland. **PreN**

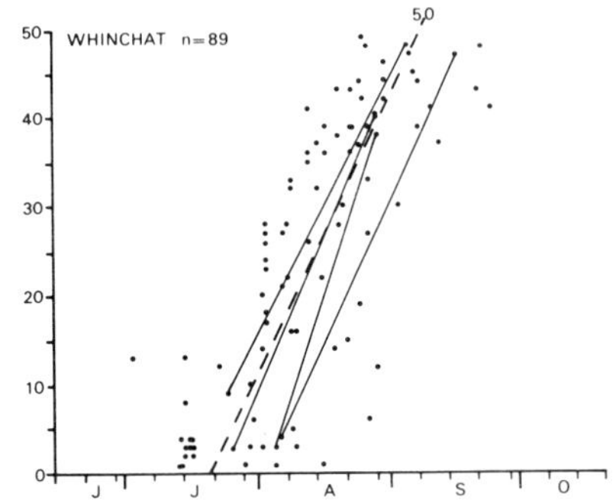

partial (B, 3-4 inner GC, exceptionally 2-3T) Feb.-March. **PJ** partial (B, 2-3 inner GC (never all), sometimes some T) June-Sept. **(237,298,327,544,710).**

STONECHAT *******S. torquata*

P:10, o.m.r. S:6,E T:3 R:12

Ad postN complete mid July/Aug.-Sept./early Oct. P duration c50 days in UK

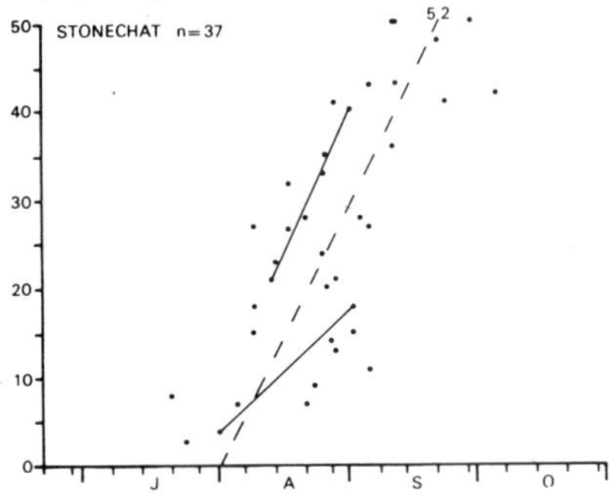

and Finland. No **PreN** moult, colour change due to abrasion. **PJ** partial (B, variable number GC, usually 3 inner but some (1st broods?) moult all and later birds may not moult any) July-Sept. **(298,492, 544,581,710).**

WHEATEAR ********Oenanthe oenanthe*

P:10, o.m.r. S:6,E T:3 R:12

Ad postN complete before migration, late June/mid July-late July/early Sept.; P duration c40 days; R start soon after P, irregular; S start when P well advanced;

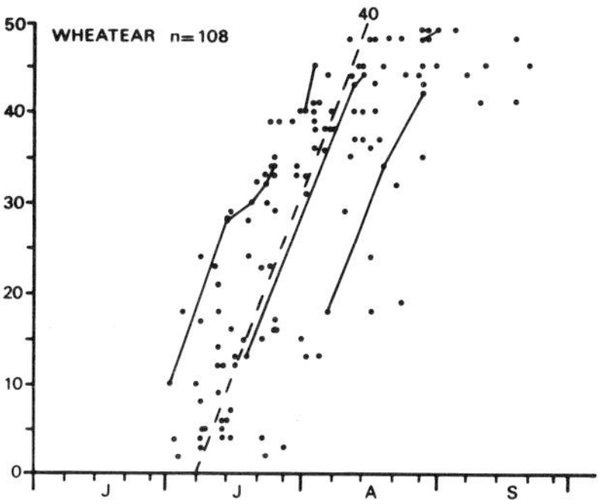

some northern *leucorrhoa* may start to migrate before P9 and inner S full grown. **PreN** partial (B, occasionally inner GC and T9) Jan.-Feb. **PJ** partial (B, sometimes some GC) starts when 5-8 weeks old, July-Sept. **(544,695,710).**

RING OUZEL **Turdus torquatus*

P:10, o.m.r. S:6,E T:3 R:12

Ad postN complete before migration, July-Sept.; UK Pscores: 21 June 6, 23 July 9, 30 July 25 (Sscore 2, T score 6). **PJ** partial (B, some inner GC (usually 3/4/5 old retained)) July-Sept. **(710).**

BLACKBIRD *****T. merula*

P:10, o.m.r. S:6,E T:3 R:12

Ad postN complete end May/late Aug.

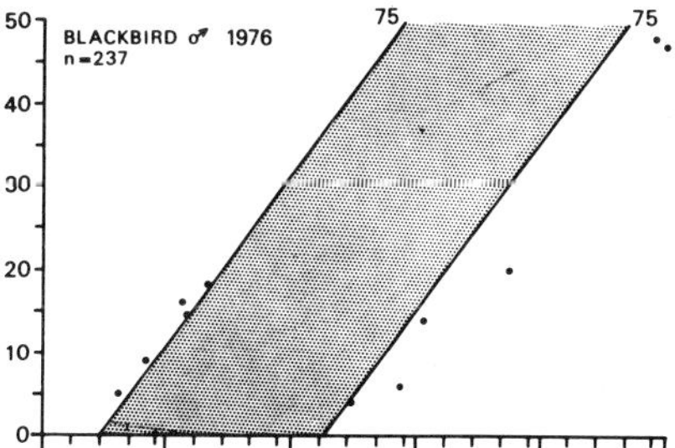

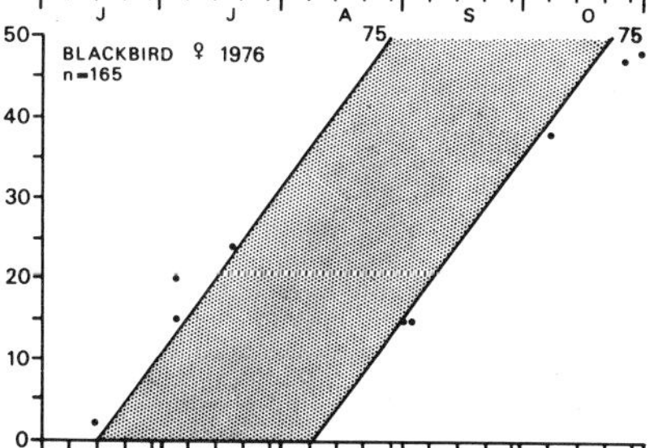

early Sept./late Oct., starting c15 days after end of breeding, no difference in timing of ♂ and ♀; P duration estimates in UK variable: Midlothian av 87 days (4 retraps 83-104 days), Cambridgeshire av 66 days (3 retraps av 117 days). **PJ** partial (B, lesser and median WC, variable number inner GC, a few also moult some T and R) July-Oct.; the extent of replacement of GC and Carpal Covert is variable: in Switzerland all birds moult at least 1 GC (most replace 6-7), 64% replace Carpal Covert 31% replace proximal Alula feather, 27% replace 1-3T; at Aberdeen only 38% replace Carpal Covert and 1% part of Alula, those which replace Carpal Covert have progressed further with GC moult, av 7 GC replaced in Aberdeen and Berkshire. **(35,207,237,298, 324,490,542,544,550,608,622).**

FIELDFARE **T. pilaris*

P:10, o.m.r. S:6,E T:3 R:12

Ad postN complete June/July-Sept.; P duration 51 days Finland. UK P,S and T scores: 17 Aug. P37, S13, T15; 18 Aug. P-, S16, T-; 19 Aug. P33, S3, T-; 20 Sept. P31, S29, T15; 22 Sept. P33, S29, T15. **PJ** partial (B, GC very variable, sometimes some R) July-Sept. **(234,237,298,327,710).**

SONG THRUSH ****T. philomelos*

P:10, o.m.r. S:6,E T:3 R:12

Ad postN complete late June/mid Aug.-mid Aug./mid Oct. P duration c50 days. **PJ**

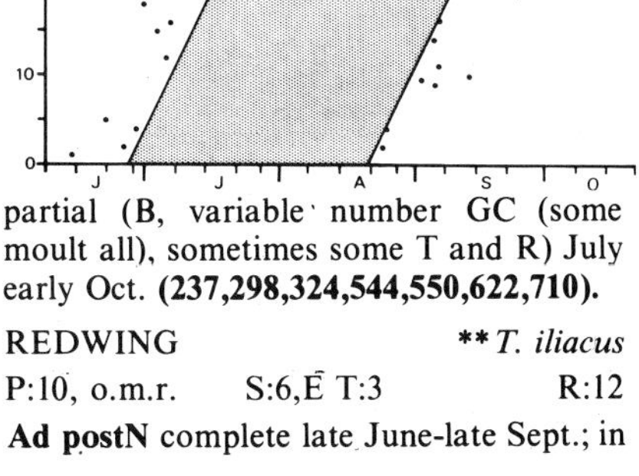

partial (B, variable number GC (some moult all), sometimes some T and R) July early Oct. **(237,298,324,544,550,622,710).**

REDWING ***T. iliacus*

P:10, o.m.r. S:6,E T:3 R:12

Ad postN complete late June-late Sept.; in

Finland some Ad start P whilst still feeding young; P duration Finland c52 days at 61° N and c40 days at 70° N. **PJ** partial (B, variable number GC (some moult all GC) and T) July-Sept. **(234,237,327,492,710).**

MISTLE THRUSH ****T. viscivorus*

P:10, o.m.r. S:6,E T:3 R:12

Ad postN complete late May/late June-late Aug./late Sept. P duration c90 days; some

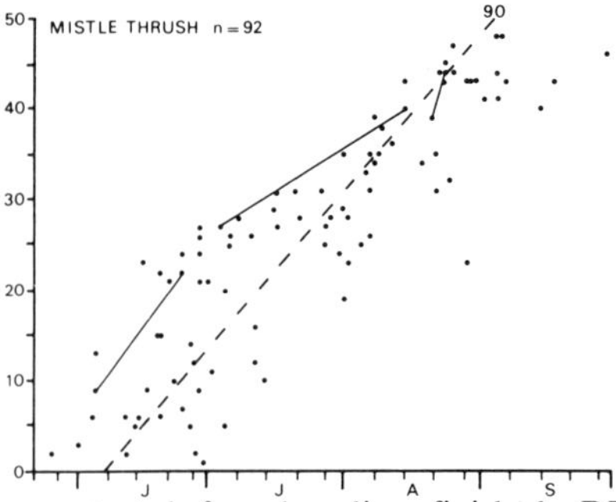

may start before breeding finished. **PJ** partial (B, usually 3-6 GC replaced — exceptionally all or none) May-Aug. **(544,608,710).**

SYLVIIDAE

CETTI'S WARBLER ***Cettia cetti*

P:10, o.r. S:6,E T:3 R:10

Ad postN complete July-Sept. (Oct?), ♂ may be earlier than ♀; R simultaneous?; S start when P5 dropped; T irregular; UK Pscores: 26 July 8; 8 Aug. 12; 9 Aug. 15; 15 Aug. 7; 3 Sept. 39; 4 Sept. 18,19; 7 Sept. 36; 12 Sept. 37; 25 Sept. 49,49. Some Ad undergo partial **preN** moult (B,WC,T?) March/April, R variable and irregular (in UK most of those replacing R appear to start at centre); the extent and pattern of Ad preN require further study. **PJ** partial (B, and at least sometimes inner 2 T) mid July-mid Sept. **(173,592,702,710).**

GRASSHOPPER WARBLER
**Locustella naevia*

P:10, o.r. S:6,E T:3 R:12

What little information there is on moult in this species is somewhat confusing. A complete **Ad postN** moult Aug.-Sept. and probably another complete **preN** moult Feb.-March noted **(710),** but it is more likely that there is one complete moult in winter quarters **(702).** There appears to be a partial moult in West Europe before migration (B, lesser and median WC,T, sometimes centre R), however a few may start P: one Norfolk 16 Aug. replacing P1 and P2 (score 7), and one South Germany 1 Sept. Pscore 12. **PJ** partial (B sometimes some inner GC) before migration, complete moult in winter quarters? Further study required. **(298,543,608,702,710).**

SAVI'S WARBLER **L. luscinioides*

P:10, o.m. S:6,E T:3 R:12

Ad postN complex and irregular; Ad migrating through Iberia in autumn with all old, all new or suspended P; a small number moult P in normal descending sequence, most moult 'eccentrically' starting from middle of P tract (c20 of 25 in active moult and all 6 which had suspended **(619)**); most birds showing eccentric moult start at P4 and moult in sequence 4-5-3-2-6-7-1-8-9-10; S start late and appear to be shed almost simultaneously; T8 is dropped first at about same time as P2 or P6; R rapid; **PJ** partial (only B?) July-Sept. **(359,562,584,608,619, 634,643,690,702).**

SEDGE WARBLER
**Acrocephalus schoenobaenus*

P:10, o.m. S:6,E T:3 R:12

Ad postN partial (B, sometimes T and some R) before migration (exceptionally whilst on passage). Complete moult in winter quarters but details poorly known: moult in West Africa c Oct.-Dec., most already in new plumage on arrival in Uganda in Dec.; Kenya in Dec. either all new or all old plumage (2 birds with suspended P moult) the latter moving further South to moult, 1

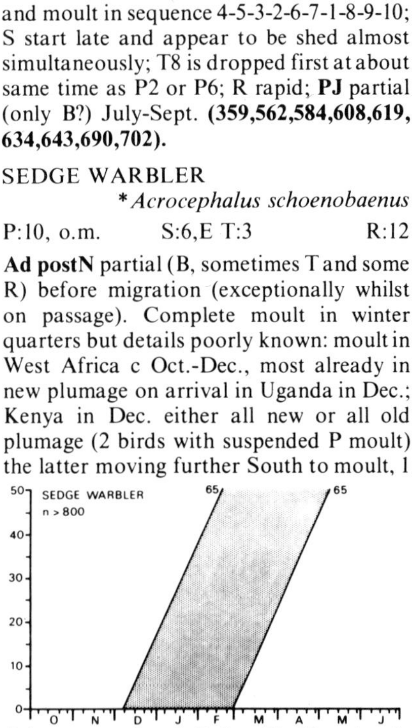

Primary moult of Sedge Warblers in the Lusaka area of Zambia 1971-75: data provided by J. J. Tucker and D. M. Francis

with P8-P10 growing 4 Feb. South West Africa; in Zimbabwe c 50% moult before, the rest after arrival; in South Africa and Botswana most appear to moult after arrival. South Malawi variable — 3 in fresh plumage late Dec./early Jan., 1 bird started moult late Nov. but most started or resumed after suspension (usually P1-P3/P4 new) in Jan., many finished by late March, est P duration 2 retraps 59 and 75 days. P moult in Zambia mid Dec.-end April, duration 65 days (J. J. Tucker and D. M. Francis pers. comm.). **PJ** partial (B) before migration, complete moult in winter quarters. **(53, 222, 298, 440, 441, 448, 452, 543, 638, 650, 702).**

MARSH WARBLER **A. palustris*

P:10, o.m. S:6,E T:3 R:12

Ad and **Juv** B moult before migration, from early July; Ad start before end of breeding. Ad and Juv complete moult in winter quarters, late Dec./Jan-April, est P duration 60-80 days. **(55,222,298,442, 447,702).**

REED WARBLER ****A. scirpaceus*

P:10, o.m. S:6,E T:3 R:12

Ad postN partial (B, 2 of 130 had replaced 2R in Hampshire) mid July-Sept. **PJ** partial (B sometimes some inner GC) starts as soon as Juv plumage fully grown; earliest birds finish last week July (Hertfordshire) but some still in Juv plumage early Sept. A very few Ad undergo complete moult in Spain in autumn. **PreN** Ad and Juv complete in winter quarters; Oct.-Dec. in West Africa, late Dec.-early March in Uganda, av P duration 70-80 days but considerable variation; est P duration 1 retrap Malawi 82 days; of 2 fresh plumaged birds in South Malawi in April, 1 showed signs of having previously suspended after renewing P1-P4. There is much partial, and usually irregular R moult on breeding grounds, especially in newly arriving birds and in early part of breeding season. **(222,272,440,441,452,552, 638,702).**

DARTFORD WARBLER ***Sylvia undata*

P:10, o.r. S:6,E T:3 R:12

Ad postN complete Aug.-Nov.; UK Pscores: 3 ♂ 11-16 Sept. 20,23,26; ♂ 1 Oct. 48, ♀ 27 Sept. 46. **PreN** partial (B, some R?) Feb.-March. **PJ** partial (B, T (irregular), some/all R?) Aug.-Nov. **(61,277,608,703, 710).**

LESSER WHITETHROAT ****S. curruca*

P:10, o.r. S:6,E T:3 R:12

Ad postN complete before migration early July-late Aug./Sept.; P duration c 40 days;

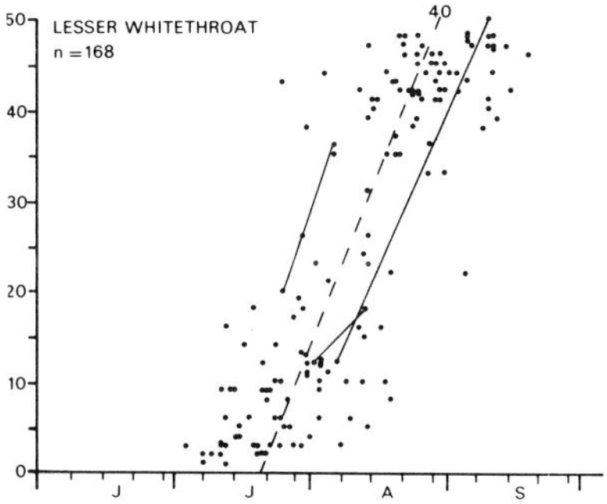

Ad apparently migrate soon after finishing P moult but not before; in UK 1 Ad with recently fledged young Pscore 2 on 21 July. **PreN** partial (B,4 inner S, some R) in winter quarters in Sudan 45% moulting R March-April (unaged). **PJ** partial (B, up to 6 GC (occasionally all?)), June-Sept. **(237,277, 298,344,441,552,703).**

COMMON WHITETHROAT *****S communis*

P:10, o.r. S:6,E T:3 R:12

Ad postN complete early July/mid Aug.-early Aug./late Sept.; P duration c40-45 days; late moulting birds may not finish before migrating (a few may not even start); some birds on passage with growing S;

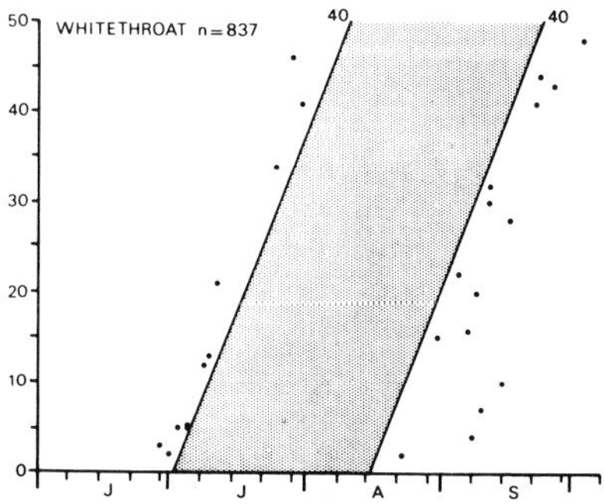

small numbers with suspended P, (3% in Iberia, 8% in Crete) with variable number old P retained (apparently more frequent in eastern *icterops* in Kenya); more birds with all new P and suspended S (3% Iberia, 67% Crete); S start when Pscore 20-30, S duration 30 days, sequence rather irregular, S6 apparently moulted with T; R start and finish before S. **PreN** partial (B, inner S, some R) or complete (apparently usually partial in *communis)* March/April. **PJ** partial (B, T and sometimes S6, up to c6 GC (occasionally all?)), late June-Sept. **(42, 73, 144, 233, 236, 237, 247, 277, 298, 327, 359, 441, 447, 470, 493, 542, 552, 582, 612, 703, 710).**

GARDEN WARBLER *******S. borin*

P:10, o.m. S:6,E T:3 R:12

Ad postN variable; most partial (B), but a few undergo complete moult; in UK there are 14 records of active P moult and a further 13 with active T but all old P (occasional birds in active P moult also found in Sweden, Belgium, Netherlands, Spain); some, at least, suspend P, 1 in UK 11 Aug. suspended after replacing P1-P4, 3 of 74 in Iberia suspended (2 birds replaced P1, 1 bird P1-P4); in UK the few which P moult apparently start early July, thus Ad with all new P unlikely to be found before c10 Aug. **PreN** complete in winter quarters late Nov.-March, P duration 65-75 days; S start at Pscore 18-25, finish about same time as P; R start at Pscore 20-25, finish at Pscore 35-40; head moult finishes 5-10 days after P; not known whether birds which undergo complete moult in autumn also undergo

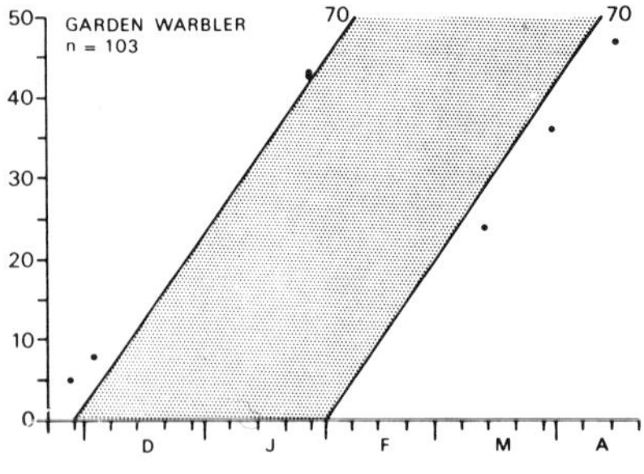

Primary moult of Garden Warblers caught at Kampala, Uganda: redrawn from Pearson (1973a)

complete spring moult. **PJ** partial (B, usually 6 or more GC, inner S) June-Sept., duration 30-35 days; complete moult in winter quarters starting a little later than Ad. **(62, 63, 78, 124, 136, 192, 222, 247, 277, 298, 331, 359, 395, 441, 447, 552, 608, 703, 710).**

BLACKCAP ***********S. atricapilla*

P:10, o.r. S:6,E T:3 R:12

Ad postN complete before migration, early July/late Aug.-late Aug./late Sept., some early birds finish before others started; ♀

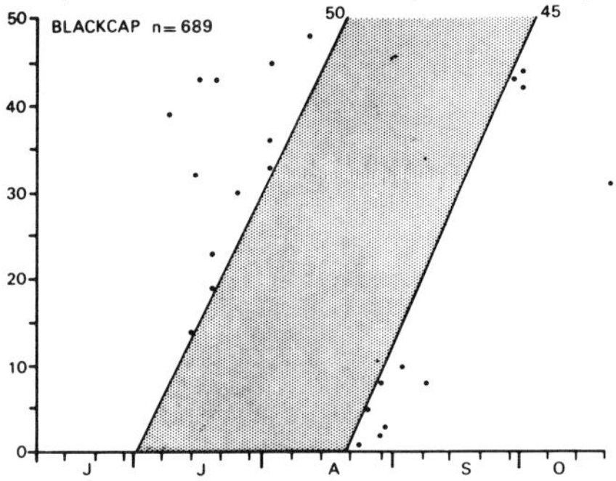

may be later starting than ♂; P duration 45-50 days, late starters apparently moult more quickly than early ones. **PreN** partial (B, 1-4 inner S (T), some/all GC) Jan.-March. **PJ** partial (B, some inner S, 6 or more (often all) GC), July-Sept. **(62, 63, 237, 277, 298, 331, 440, 441, 552, 622, 703).**

WOOD WARBLER **Phylloscopus sibilatrix*

P:10, o.m.r. S:6,E T:3 R:12

Ad postN partial (B, sometimes T and corresponding GC, lesser WC and occasionally 1 or more R), mid June-early Aug. before migration. **PreN** complete in winter quarters mid Dec./Jan.-March. **PJ** partial (B) starts when c30 days old, July-Aug. **(209,210,572,625,701).**

CHIFFCHAFF ********P. collybita*

P:10, o.r. S:6,E T:3 R:12

Ad postN complete late June/late Aug.-early Aug./early Oct. before migration; P duration c45 days. **PreN** partial, very variable in extent (B, sometimes T and centre R) Dec./Jan. **PJ** partial (B,T (occasionally S4-S7 in captive birds), some inner GC (occasionally all), some replace

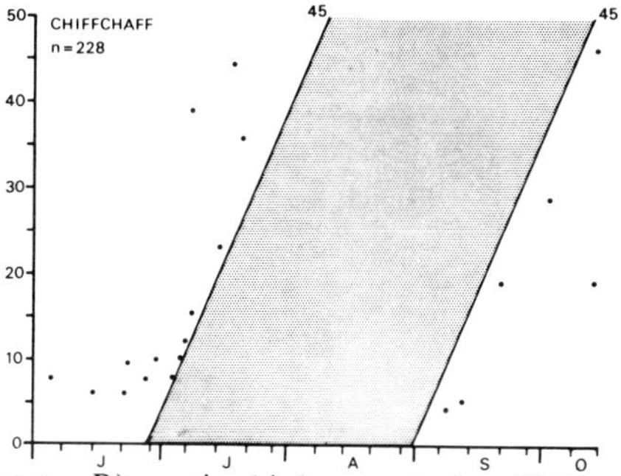

centre R); captive birds started when 60-70 days old, in wild a lapse of 2-3 weeks between completion of Juv plumage and start of PJ moult, July-Oct. **(210,211,212, 237,298,383,416,441,518,552,625,701).**

WILLOW WARBLER ****** *P. trochilus*

P:10, o.r. S:6,E T:3 R:12

Unique among British Passerines in that it undergoes 2 complete moults each year. **Ad postN** complete before migration early June/late July-late July/early Sept.; P duration c30-40 days UK, 37(29)-40 days

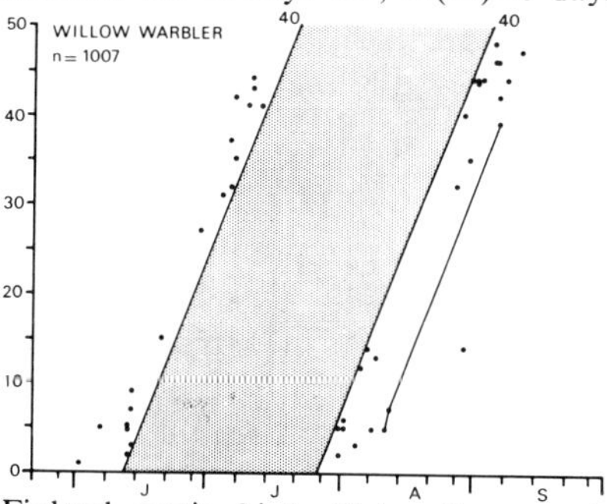

Finland, captive birds c50 days Germany; in Finnish *acredula* some ♂ may start P during incubation others not until young fledged, ♀ later (10-15 days?); in UK ♂ and ♀ may start P moult whilst feeding young, especially in late/replacement broods. In Cleveland ♂ start P moult over period of 14 days, ♀ start extends over 21 days and a little later than ♂; migration apparently starts as soon as moult finished; some suspend/arrest S (6 of 250 in Iberia, 40 of 450 in Crete); 57% in Crete did not moult S in normal sequence, S1,S2,S3 and S6 were the most likely to be replaced before suspending. **PreN** complete in winter quarters mid Dec-early April; P

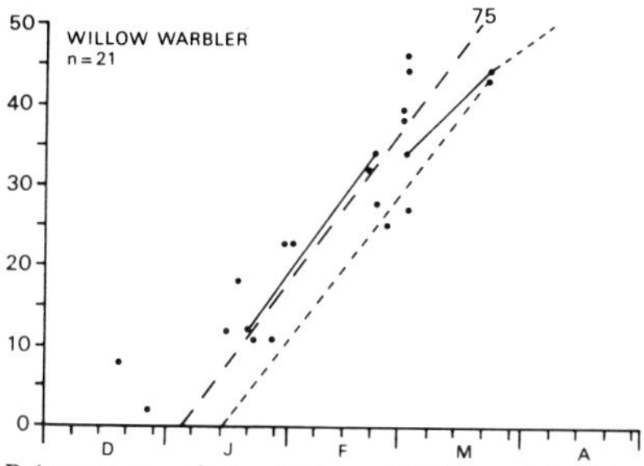

Primary moult in Africa of Willow Warblers caught at Kampala, Uganda: redrawn from Pearson (1973a)

duration est 65-80 days Uganda, 60-70 days Kenya; S start at Pscore 20-30, inner 2S finish growth after P; R start at Pscore 25, finish Pscore 35, centrifugal. **PJ** partial (B, all birds moult all lesser WC, most only moult inner median WC (a few all), a few moult 1-2 inner GC); moult starts when c23 days old late June-Sept.; apparently some birds start moving before finishing moult. Most undergo complete moult in winter quarters as Ad, but a few birds apparently only moult B and WC, having very worn P,S,R in spring. In Belgium 1 of 16 birds in May had not undergone any preN moult. **(10,25,42,53, 136, 163, 178, 209, 210, 211, 212, 214, 233, 234, 236, 237, 252, 298, 327, 359, 415, 416, 441, 447, 507, 552, 608, 612, 622, 624, 625, 701).**

GOLDCREST *** *Regulus regulus*

P:10, o.r. S:6,E T:3 R:12

Ad postN complete late June-mid Oct.; P duration c55 days, av 52 days in captive birds; some may start P moult after 1st

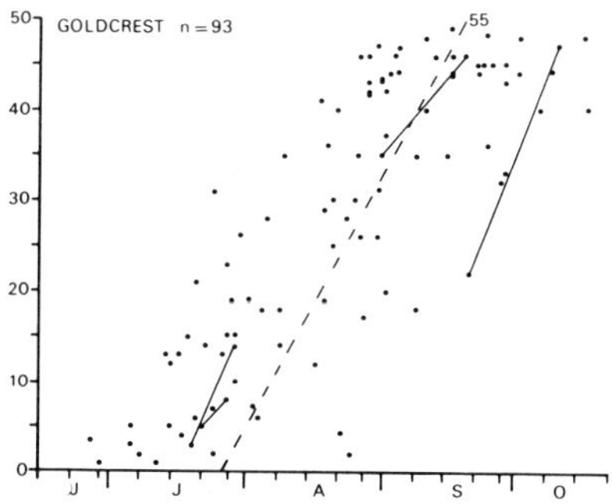

brood and suspend during 2nd brood (1 suspended after renewing P1-P3 27 July). **PJ** partial (B, median and lesser WC, sometimes some inner GC and T8) July-Sept., duration c50 days. **(298,616,710).**

FIRECREST **R. ignicapillus*

P:10, o.r. S:6,E T:3 R:12

Ad postN complete July-Sept., duration c53 days. **PJ** partial (B, median and lesser WC, GC and T?) July-Sept., starts when 37-39 days old. **(298,616,710).**

MUSCICAPIDAE

SPOTTED FLYCATCHER ****Muscicapa striata*

P:10, o.r. S:6,E T:3 R:12

Ad postN partial (B) July-Sept., some also moult T, some inner S and WC, and occasionally some P and centre R; in UK most moult some T, sequence irregular but most start with T8; there is 1 UK record of Ad in P moult (1 Aug: P1 score 3, P2 score 1, P3-P10 old), and 11 of birds replacing 1-3 S (irregular sequence); in Finland 63-85% moult T (sequence variable) and 1 bird moulted inner S; in Crete 3 of 119 had suspended S and 104 suspended T, in Iberia 9 of 32 had suspended P or S. **PreN** complete in winter quarters Nov.-March but some birds may suspend and finish on breeding grounds; although generally noted that P,S, and R moult is in reversed sequence (P ascendant, S descendant, R centripetal) there is considerable irregularity **(704):** P9-P7 are moulted 1st ascendantly, P10 often seems to moult later; moult of P coverts does not seem to be linked with corresponding P; Alula usually moult concurrently with P7-P10; R usually centripetal, does not appear to be in phase with any particular stage of moult; T moulted before S; S often begin descendantly from S6, but perhaps as frequently ascendantly from S1, irrespective of whether S1 or S6 dropped 1st, moult apparently occurs from both ends of S tract to meet at S3/S4. **PJ** partial (B, T and occasionally S6, some inner GC) July-Sept.; complete moult in winter quarters as Ad but rather later. **(134,136,237,269,359,587, 612,696,704,710).**

PIED FLYCATCHER ***Ficedula hypoleuca*

P:10, o.r. S:6,E T:3 R:12

Ad postN complete June/July-Aug.; in Finland all P and T replaced but variable number S and it appears that not all birds replace all S before migrating; P duration c45 days Finland; in Finland c10-25% ♂ start P moult early June, but most ♀ not until mid July; in UK both sexes regularly start P moult whilst feeding young in late and replacement broods; in Iberia 4 of 421 had suspended S (all had new P). **PreN** partial (B, 3 (4) inner S, inner GC) Jan.-Feb. in winter quarters. **PJ** partial (B, a few birds a few GC) June-Aug. **(112,237,269,359, 710).**

TIMALIIDAE

BEARDED TIT *****Panurus biarmicus*

P:9+1 v. S:6,E T:3 R:12

Ad postN complete, late June/mid Aug.-late Sept./early Oct.; P duration 45-55 days in Suffolk and broadly similar estimates

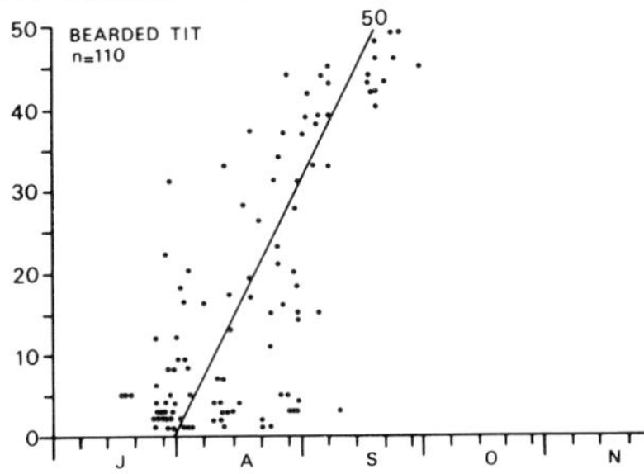

Only birds recorded as adults are plotted: additional data from Pearson (1975) added.

elsewhere; all tracts moult within period of P moult. **PJ** complete; in Suffolk most start late July/early Sept. and finish early Sept./early Oct.; early fledging birds have longer period between fledging and onset of PJ moult than later birds (those fledged by early June start PJ 7-9 weeks later, those fledged early July 5-6 weeks later, and those fledged during Aug. 2-4 weeks later); P duration apparently variable, those starting in July c55 days, in first 10 days Aug. c50 days, after 10 Aug. c44.5 days, very late starters may moult in less than 40 days; quicker moult of late birds due to more feathers being replaced at once; all birds drop P1 and P2 in quick succession, P3 (in late starters also P4) dropped before P1 starts growth; S1 shed with P4/P5; T shed in rapid succession starting when P4 dropped; R start at about same time as T; B moult unusual: moult within individual centres of tracts very synchronized; many downy feathers present in 1W birds (particularly under tail coverts and sides of dorsal tract) are lost early in breeding season. **(91,245,444,553,562).**

AEGITHALIDAE

LONG-TAILED TIT

****_Aegithalos caudatus_

P:10, o.r. S:6,E T:3 R:12

Centre R in Juv = R5, in Ad =R3

Ad postN complete late May/early July-late Aug./early Oct. P duration c80 days; S

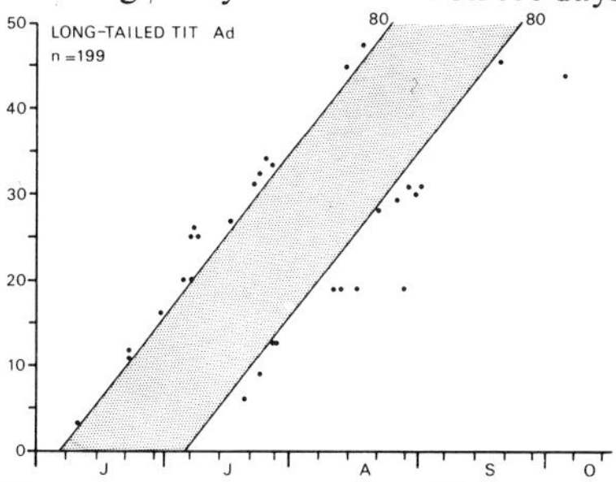

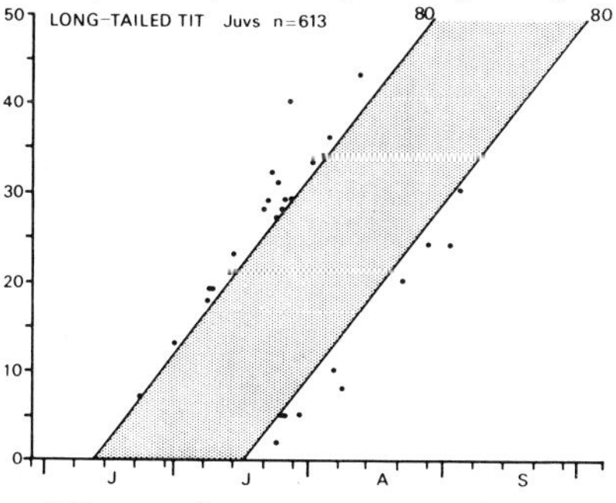

and R start c2 weeks after P at Pscore c11. **PJ** complete early June/mid July-early Sept./early Oct. **(23,217,218,356).**

PARIDAE

MARSH TIT ***_Parus palustris_

P:10, o.r. S:6,E T:3 R:12

Ad postN complete mid May/late June-late

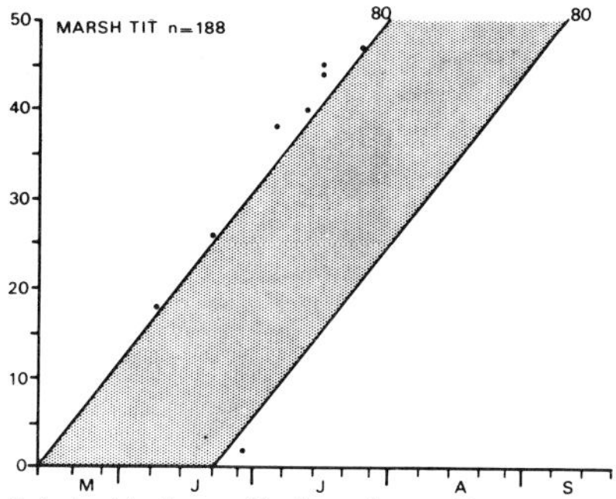

July/mid Sept. P duration c80 days. **PJ** partial (B, some/all GC, T8, some centre R) June-Sept. **(126,298,710).**

WILLOW TIT ***_P. montanus_

P:10, o.r. S:6,E T:3 R:12

Ad postN complete mid May/late June-late July/mid Sept.; P duration c70 days in UK,

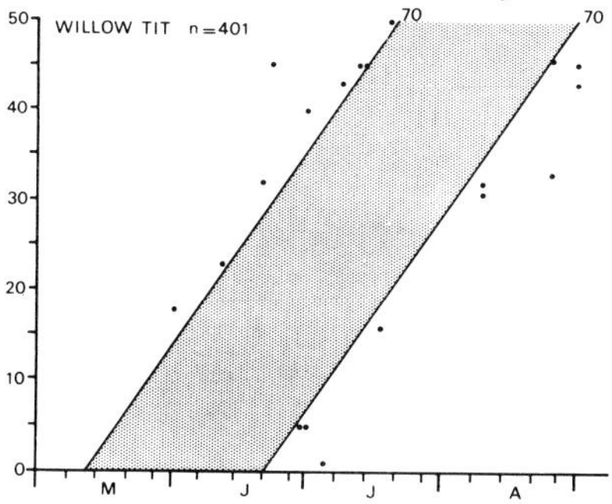

c77 days in Finland where ♂ start a little earlier than ♀ and onset of moult overlaps end of breeding. **PJ** partial (B, some/all GC, T8, some centre R) June-Sept. **(298,327, 424,622).**

CRESTED TIT *_P. cristatus_

P:10, o.r. S:6,E T:3 R:12

No information available for Scottish birds. **Ad postN** complete Aug.-Oct. (from July in USSR); no overlap of breeding and moult in Finland; in Iberia pattern, timing and duration similar to other _Parus_ spp. **PJ** partial (B, c6 GC, T8); in USSR 1st broods start July, 2nd broods Aug. **(60,126,424).**

COAL TIT ***_P. ater_

P:10, o.r. S:6,E T:3 R:12

Ad postN complete, late May/late June-mid Aug./early Sept. P duration c70 days;

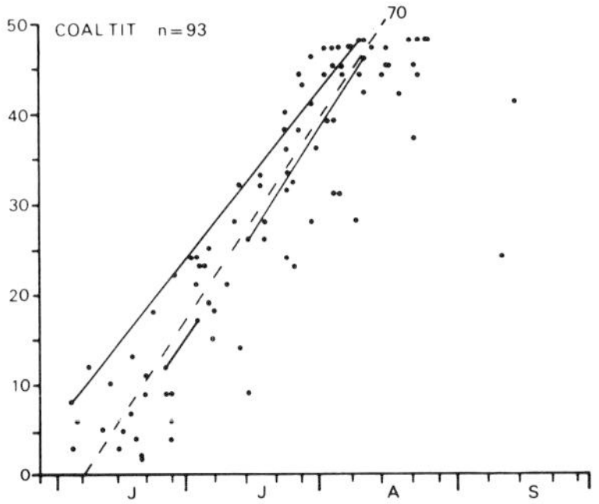

PJ partial (B, 4-7 GC in 1st broods (2nd broods may retain all), 1-3T); timing and extent variable; in UK a very very few moult all GC mid July-mid Sept.; in French Alps 30% renew Alula. **(298,512,550).**

BLUE TIT *****P. caeruleus*

P:10, o.r. S:6,E T:3 R:12

Ad postN complete and commonly starts a few days before young fledge; mid May/early July-early Aug./late Sept.; ♂ sometimes start before ♀, and 1S before older birds; in England Southern breeders start somewhat earlier than Northern breeders, the latter moult more quickly; P duration 62-85 days Cambridgeshire; data

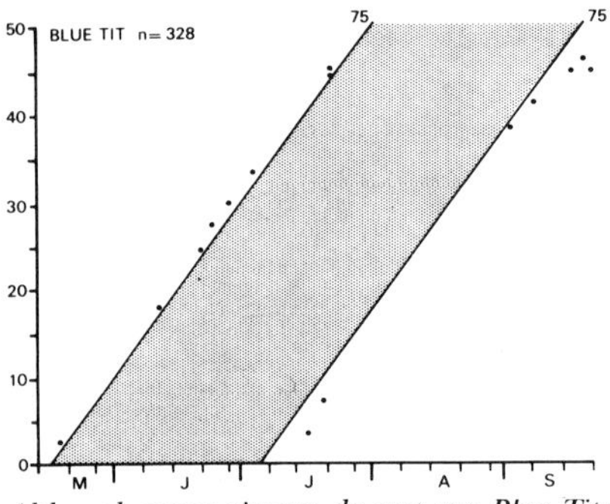

Although many ringers do not sex Blue Tits only those recorded as males are plotted

for all England total moult 115-120 days, P moult covers c1st 60% of this period, ♂ 75 days, ♀ 80 days; P duration c70 days Finland; S start at Pscore 10-30 and is rarely finished by end P moult; B and head continue for c1 month after P replaced; R start soon after P (most start before Pscore 15) and finish from Pscore 30. **PJ** partial (B, 1-3 T) mid July-early Oct.; extent of moult varies from year to year and between sexes, sometimes most retain 2-3 (sometimes up to 6) old GC, in other years few have any old GC and up to ⅓ also renew Alula and Carpal Covert; c15% renew centre R. **(136, 171,180,237,298,327,453,550,608,622).**

GREAT TIT *****P. major*

P:10, o.r. S:6,E T:3 R:12

Ad postN complete, may start before young fledge; late May/late July-late July/early Oct.; ♂ sometimes start before ♀; in England Southern breeders start earlier than Northern birds; P duration ♂ 75 days,

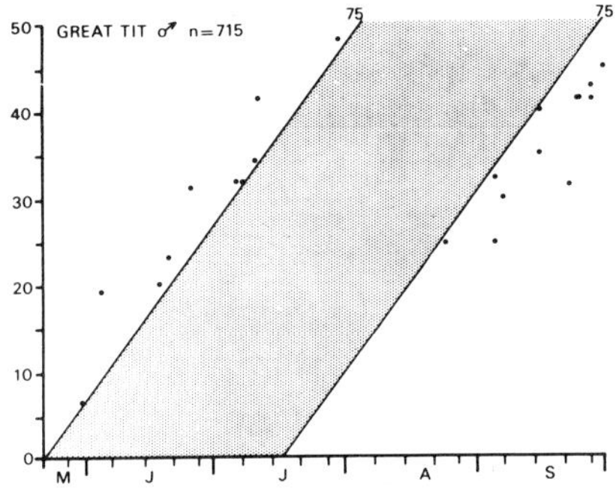

♀ 75-80 days late starting birds being slightly quicker in UK, c70 days in Finland, in Belgium c89 days (captive birds 86-99 days); S start at Pscore 10-25 and may continue for several days after P finish; R start at same time as S, finish Pscore 45; T start Pscore 10-25, finish Pscore 40. **PJ** partial (B, 1-3T) late July-late Sept.; in Kent 80% replaced all R but none replaced Alula; 8% in Hertfordshire replace all/part Alula; extent of GC moult variable, often all are replaced (in Hertfordshire c20% retain some old GC-usually 3 or less); in Belgium 50+% moult Alula (twice as many ♂ moult as ♀), more early fledged birds moulting than later ones (this high proportion of birds replacing Alula is of interest since more 2nd broods in Belgium than in UK, thus one would expect more birds to moult in UK than in Belgium); 1st broods in Belgium moult R when c80 days old, 2nd broods when c65 days old, and in USSR period between fledging and start of PJ moult shorter in later fledged birds, later fledged birds also finish moult more quickly; in UK usually moult some centre R seldom all, in Belgium and Germany moult all R. **(129,130,136,171,237,298,327,424, 550,608).**

SITTIDAE

NUTHATCH **Sitta europaea*

P:10, o.r. S:6,E T:3 R:12

Ad postN complete, late May-mid Sept. P duration c80 days; evidence for a **preN**

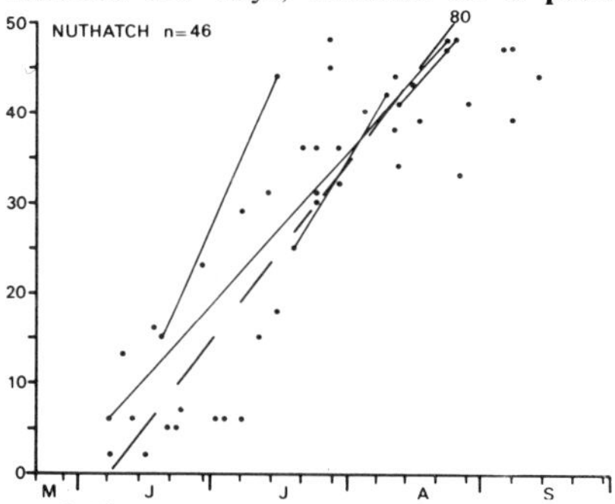

moult "is very scanty and is not conclusive" **(49)**. **PJ** partial (B, but not GC? some T?) July-end Aug. **(49,710).**

CERTHIIDAE

TREECREEPER ***Certhia familiaris*

P:10, o.r. S:6,E T:3 R:12

Ad postN complete early June/early July-late July/early Sept.; P duration c70 days; S

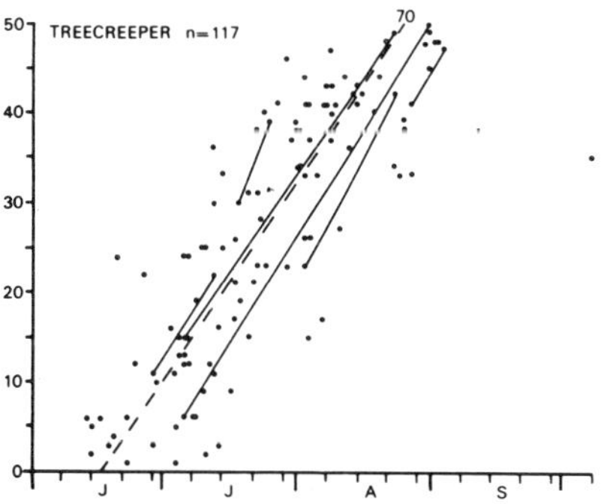

start at Pscore c20, finish about same time as P; R2-R6 dropped almost simultaneously soon after P start, once outer R sufficiently well grown to provide support R1 dropped, R1 finish growth 1-2 weeks after rest of R and at about same time as P finish. **PJ** partial (B, all GC, some inner S) July-Oct. **(170,241,581,710).**

ORIOLIDAE

GOLDEN ORIOLE **Oriolus oriolus*

P10 S:6,E T:4/5 R:12

Ad postN partial (B, some birds also replace some inner S and some central R) before migration, ♀ starting late June/July; 4 of 11 in Iberia had suspended S, all had old P. **PreN** complete in winter quarters Nov.-early Feb.; in captive birds total duration c70 days, S1 dropped 1st, then S2 and S7, then S8 and S9 (captive); R centrifugal. **PJ** partial (B) starts soon after fledging; complete moult in winter quarters as Ad. **(126,298,359,608,718).**

LANIIDAE

RED-BACKED SHRIKE **Lanius collurio*

P:10, o.r. S:6,E T:4 R:12

Ad postN partial (B) before migration, early July-Sept., with a peak in late July/early Aug.; a few start wing moult; of 130 in Crete, 4 suspended P (2 birds had replaced P1, 1 bird P1 and P2, 1 bird P1-P5), 1 had suspended S, 82 had suspended T, more birds in suspended moult in early Sept. than later in month. **PreN** complete moult in winter quarters, from Oct. in Zaire, Dec.-March in East Africa; flight feathers replaced in 80-85 days. **PJ** partial (some B, c5 (occasionally more, never all) GC) starting immediately after fledging; complete moult in winter quarters Nov.-Feb. **(53, 126, 215, 276, 298, 318, 440, 447, 542,608,612,650).**

CORVIDAE

For details of pterylosis of European species see Morlion and Vanparijs (1979). Seel (1976) details moult in all feather tracts.

JAY ***Garrulus glandarius*

P:10, o.r. S:6,E T:4 R:12

Ad postN complete early June/mid July-early Sept./mid Oct.; ♂ and ♀ at same time; complete moult lasts c102 days, P duration c92 days, S duration c74 days, T duration c53 days, R duration c66 days; moult starts with dorsal tract, P1 drops c5 days later, S1 drops with P4/P5; moult continues in ventral and dorsal tracts throughout winter Nov.-March. **PJ** partial (B, lesser WC,

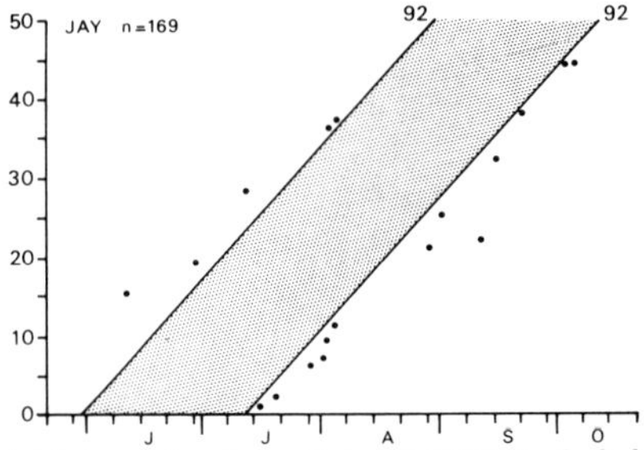

Additional data from Holyoak (1974) included

sometimes 1T), moult of GC variable, in UK c6 or less, in Switzerland 50% moult 1 or more (3 of 50 replaced all), in Sweden none; no Swedish birds moult Alula, 50% in Switzerland moult part (3 of 50 replaced all); some may replace P1-P2 and centre R during 1st year. **(28,34,260,352,489,528, 608).**

MAGPIE **Pica pica*

P:10, o.r. S:6,E T:4 R:12

Ad postN complete early June/mid July-Sept./Oct. (Nov.); 1st birds to start moult are 2Y ♂ next 2Y ♀ and older ♂, finally older ♀ (most 2Y do not breed); P duration c95 days **(260),** 131 days **(528);** S1 starts

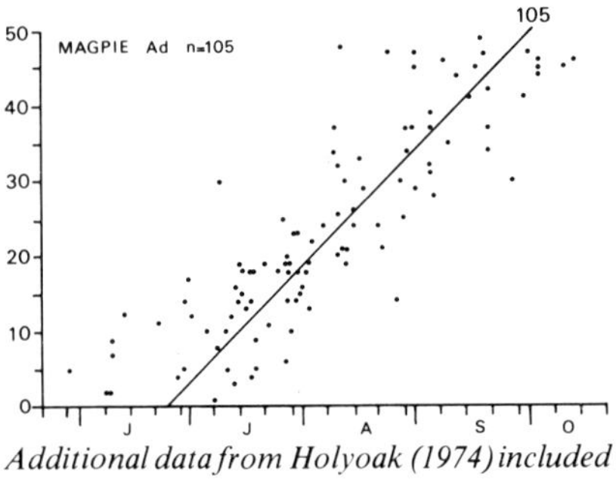

Additional data from Holyoak (1974) included

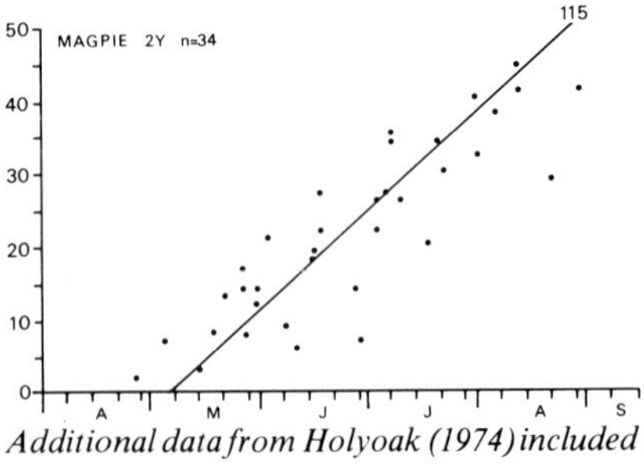

Additional data from Holyoak (1974) included

growth c46 days after P1 (apparently 2Y start S moult at a later stage than Ad breeders), S duration c113 days, T duration c61 days, R duration c88 days. **PJ** partial (B, all GC, sometimes some centre R) June-Sept. **(28,237,260,298,528,710).**

CHOUGH **Pyrrhocorax pyrrhocorax*

P:10, o.r. S:6,E T:4 R:12

Ad postN complete June-Oct.; P duration c92 days. **PJ** partial (B, lesser and median WC) Aug.-Sept. **(260,710).**

JACKDAW ***Corvus monedula*

P:10, o.r. S:6,E T:4 R:12

Ad postN complete late May/early July-mid Aug./early Oct.; 2Y start before older birds; P duration c105-127 days (c166 days in captive birds); S1 starts growth 48 days after P1, S duration c94 days, T duration c69 days, R duration c58 days. **PJ** partial (B, 3 inner GC) in autumn. **(28,260,528,710).**

ROOK ****C. frugilegus*

P:10, o.r. S:6,E T:5 R:12

Ad postN complete early May/late June-early Aug./Oct.; 2Y birds start 1st, next older ♂, finally older ♀ (which may moult more quickly); Ad in captivity dropped P1 2-3 days after young left nest; P duration

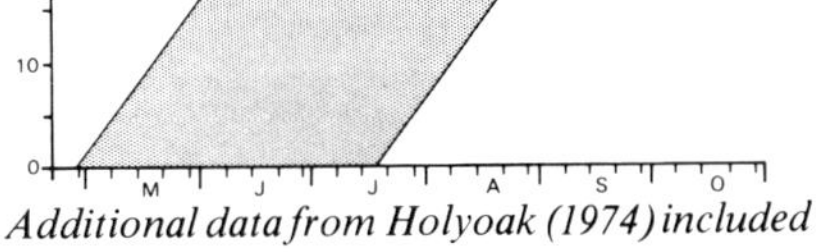

Additional data from Holyoak (1974) included

c107 days **(260),** c162 days **(528);** S1 drops c45 days after P1, S duration c105 days, T duration c88 days (sequence 8-9-10-11-7), R duration c92 days. **PJ** partial (B, inner 3 GC) June-Sept.; loss of feathers from chin and cheeks may start Jan.-Feb., feathers over nares later (March/April/May), in Scotland captive ♀ finished facial moult in 71 days; some birds may replace P1-P2 and centre R during 1st year. **(30,149,260, 298,339,489,528,608,709).**

CARRION/HOODED CROW ****C. corone**

P:10, o.r. S:6,E T:5 R:12

Ad postN complete early May/mid July-early Aug./mid Oct.; non-breeding 2Y start c1-2 weeks before breeding birds; P duration c133 days **(260),** c172 days **(528);** S duration c125 days; S and T in sequence 8-1-9-2-7/3-10/11-4-5-6; T duration c89 days, R duration c98 days (centrifugal); in Germany 3 of 1000 had not renewed P9 and P10. **PJ** partial (B) June-Sept.; GC variable, none-all, in Germany 35% moulted some. **(28,260,292,298,528).**

RAVEN **C. corax*

P:10, o.r. S:6,E T:5 R:12

Ad postN complete March/April-Sept./Oct.; P duration c140 days UK, 140-145 days Finland, captive birds in Germany 140-145 days; S start when P4/P5 dropped, finish at same time as P. **PJ** partial (B, some lesser and median WC?) autumn. **(208, 260,327,710).**

STURNIDAE

STARLING *****Sturnus vulgaris*

P:10, o.m.r. S:6,E T:3 R:12

Ad postN and **PJ** complete; in UK no obvious difference in timing in ♂ and ♀; Ad in South England and Wales late May/late June-early Sept./early Oct., P duration c80

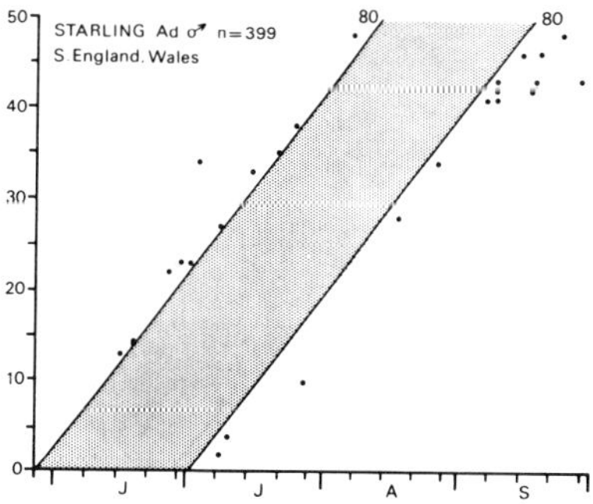

days, Ad in North England and Scotland start and finish a little later. Juvs in South England and Wales early June/mid July-early Sept./mid Oct., in North England and Scotland c1 week later, those on Scottish Isles (Fair Isle, North Ronaldsay, Lewis) start late June/early July; S1 dropped with

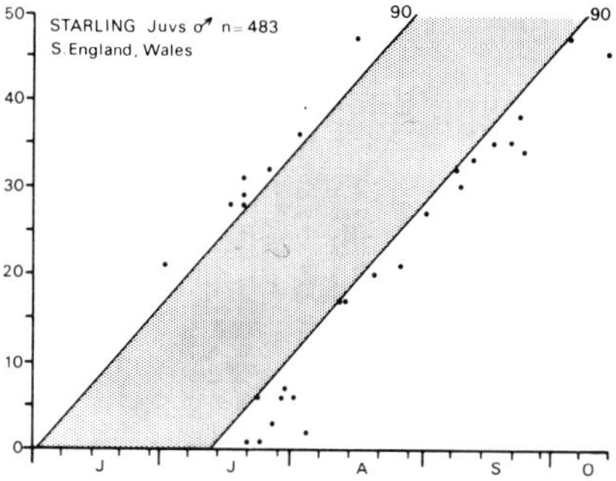

P6, S5/S6 finish at same time as P9 (P10 earlier); T, GC and lesser WC moult at same time as P1; Alula renewed when P9 ½ grown; R start late, R1 dropped 1st, then irregular; exceptionally Juv Alula and up to 3 Juv S may be retained. **(29,33,92,213,237, 298,319,388, 526,608,698).**

PASSERIDAE

HOUSE SPARROW *****Passer domesticus*

P:10, o.m.r. S:6,E T:3 R:12

Ad postN and **PJ** complete; in South England and Wales Ad ♂ mid July/late Aug.-mid Sept./late Oct., P duration c60

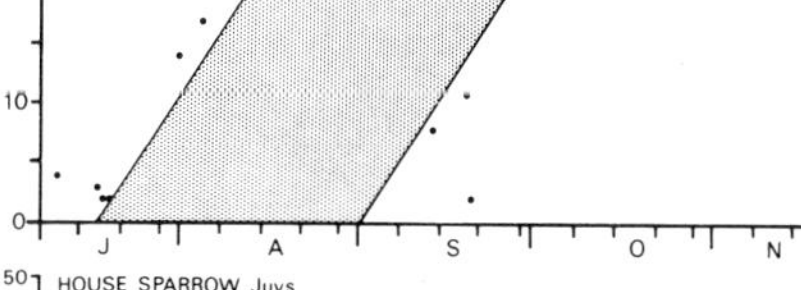

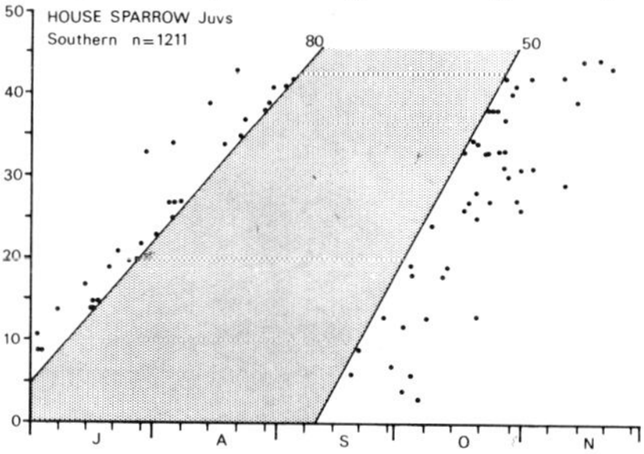

days ♀ have a wider scatter, some starting late June, some finishing early Nov.; North England and Scotland similar, ♂ a little earlier than ♀. Southern Juv mid June/early Sept.-early Sept./late Oct. early starters P duration c80 days, later birds c50 days, northern birds probably start a little later, few data; in USA PJ start c5 weeks after fledging. In Germany total moult duration of Ad 82 days, Juv quicker; P duration 65-85 days Finland. S start when P5 dropped, sequence 8-1-9-7-2-3-4-5-6; Alula starts when P7 dropped, GC moulted simultaneously with P4; R centrifugal starting when P5 dropped, finish while P9 growing; B starts when P3 dropped. In Texas 6 of 130 Ad had suspended after replacing P1 and P2. **(31,69,96,188,298,327, 681,717).**

TREE SPARROW *****P. montanus*

P:10,o.m.r. S:6,E T:3 R:12

Ad postN and **PJ** complete. In UK ad start P from late June, most starting 2nd ½ July, last birds start early/mid Sept. Juv start from mid June, most in mid Aug. (6-13 days later than Ad), but there may be a smaller earlier peak in onset of moult at same time as Ad; period between fleding and start of

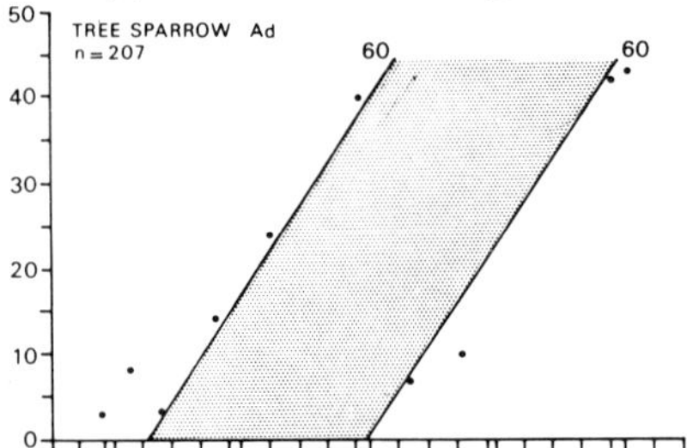

Redrawn from Bibby (1977)

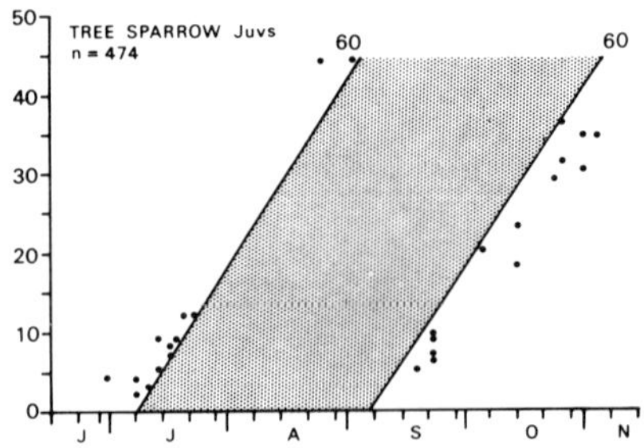

Redrawn from Bibby (1977)

PJ moult shorter in late fledged birds (c36 days post fledging 1st broods, 24-26 days 2nd broods, Hertfordshire), in Poland 2nd brood young moult more quickly than 1st broods, no difference in UK. P duration c60-70 days, Ad probably slightly quicker than Juv. S start at Pscore c22, finish at same time as P or up to 17 days later, S duration c53 days. In Austria 1 Ad feeding young having replaced P1 and suspended. **(65,66,103,123,295,396,472,675).**

FRINGILLIDAE

CHAFFINCH *****Fringilla coelebs*

P:9+1 m. S:6,E T:3 R:12

Ad postN complete, early June/late July-mid Aug./late Sept. (early Oct.); P duration c70 days UK, c70 days Finland, in USSR 65

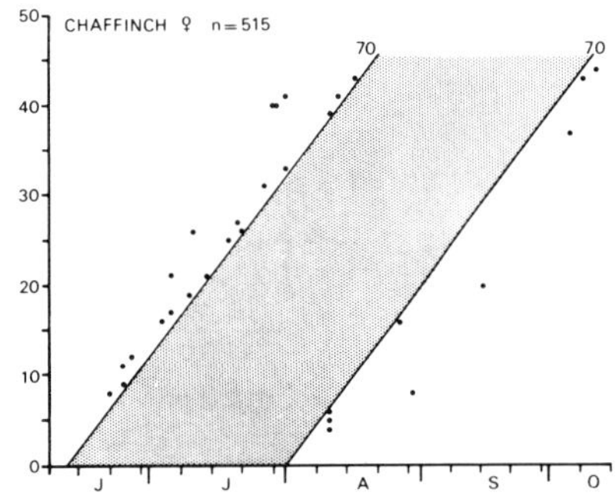

days at 60° N and 70-80 days in Crimea/Ukraine; 'arrested' moult recorded in migrant birds in 2nd ½ Sept./Oct. in White Sea/Baltic Sea area. **PJ** partial (B, c5+ GC (often all), 1-3 T) early Aug.-late Sept.; in Baltic PJ starts 20-30 days after completion of Juv plumage; duration PJ moult 45 days at 60° N in USSR, 50 days in Crimea/Ukraine. **(136,137,138,139,188, 235,237,327,405,417,551,623).**

BRAMBLING **F. montifringilla*

P:9+1 m. S:6,E T:3 R:12

Ad postN complete before migration, in Finland mid July-early Sept.; ♂ start c2 weeks before ♀ while most of population has nestlings but not known whether ♂ start moult whilst feeding young; P duration c55-58 days Finland; Ad ♀ Scotland 27 July Pscore 23. **PJ** partial (B, some GC) Aug.-Sept. **(137,163,234,327, 608,710).**

SERIN *Serinus serinus*

P:9+1 m. S:6,E T:3 R:12

Ad postN complete late Aug.-mid Oct; Ad ♂ Lincolnshire 12 Aug. Pscore 14. **PJ** partial (B, some GC and inner S) late Aug.-mid Oct., some have complete moult in Portugal. **(298,495,608,710).**

GREENFINCH ****Carduelis chloris*

P:9+1m. S:6,E T:3 R:12

Ad postN complete early July/mid Aug. late Sept./mid Nov.; P duration c85 days UK, 75-85 days Finland, captive birds in

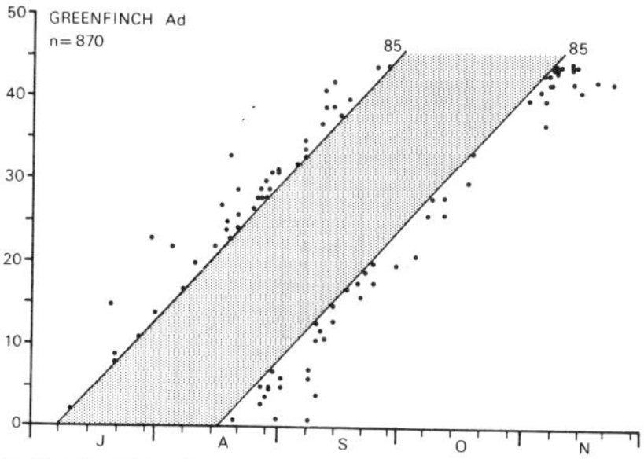

UK 90-123 days; total moult duration c105 days UK, captive birds 109-131 days. **PJ** partial (B, some inner GC, 1-2 T) late Aug.-mid Oct.; extent of PJ moult variable, 1% in UK may have complete moult, 10-20% in Portugal; 10% recorded moulting P and S in Germany; R moult variable, some in UK

Moult of Greater Coverts by Juvenile Greenfinches Carduelis chloris *in North Nottinghamshire (M. Boddy unpubl.) % of birds retaining old Greater Coverts.*

	No. old GC retained								
	0	1	2	3	4	5-7	8-9	N	
♂	44	19	11	11	3	12	0	73	1979/80
♀	47	11	19	9	1	12	1	70	
♂	55	17	9	5	5	8	1	398	1981/82
♀	43	11	18	5	4	16	3	247	

All birds caught between mid Nov. and mid May.

replace all R, in Finland c15% replace all R and 20-25% replace up to 5R (centrifugally); a few in UK replace Alula. **(70,298,326,327, 398,404,405,493,551,608,622,623,688).**

GOLDFINCH ****C. carduelis*

P:9+1m. S:6,E T:3 R:12

Ad postN complete early July/late Aug. early Sept./late Oct. P duration c70 days;

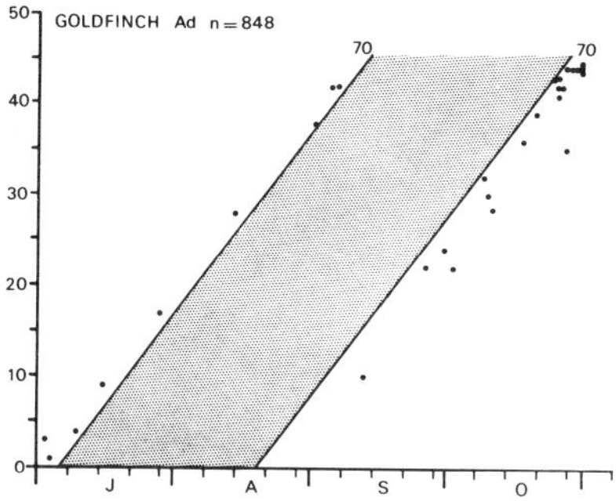

S1 drops with P5/P6; R starts when P4/P5 dropped in sequence 2-3-4-1-5-6 (Australia). In Northern Hemisphere moult starts after breeding but in Australia may start during egg laying. **PJ** partial (B, c5+ GC (not all), 1-2 T) Aug.-Sept.; some undergo complete PJ moult, 3-4% in UK, 20-30% in Portugal, most in Spain, Sicily and Greece. **(298,370,405,623,710).**

SISKIN ***C. spinus*

P:9+1 m. S:6,E T:3 R:12

Ad postN complete starting late June/early Aug., P duration 70 days captive birds. **PJ** partial (B, 1-2 T, c5+ GC (very rarely all)) Aug.-Sept. **(137,237,298,405,608,610).**

LINNET ****C. cannabina*

P:9+1 m. S:6,E T:3 R:12

Ad postN complete, early July/late Aug.-late Aug./late Oct. P duration c70 days. **PJ**

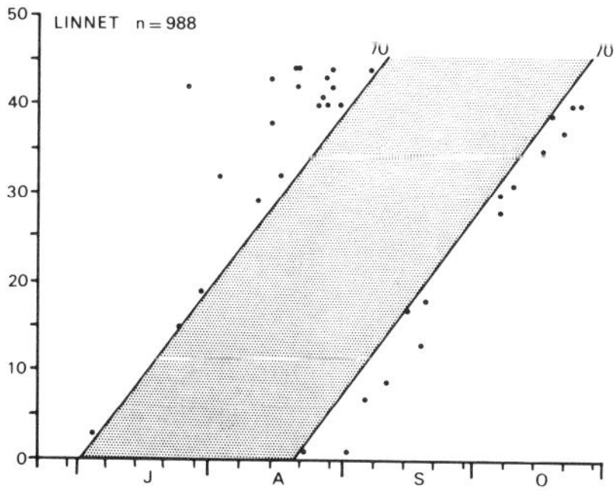

partial (B, c5+ GC (not all), 1 or a few centre R, occasionally 1-3 T) July-Oct.; complete moult very rare in UK, some regularly undergo complete moult in Mediterranean area. **(137,298,325,405,608, 623).**

TWITE ****C. flavirostris*

P:9+1 m. S:6,E T:3 R:12

Ad postN complete July-Sept./Oct.; P

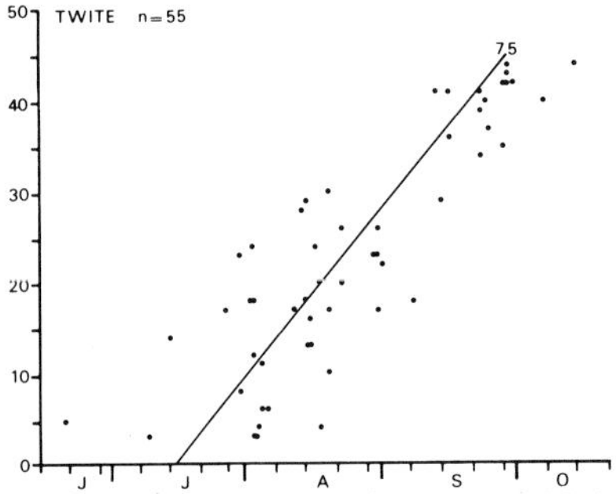

duration c75 days. **PJ** partial (B, all GC?, 1-2 T) July-Sept. **(405).**

COMMON REDPOLL *****C flammea*

P:9+1 m. S:6,E T:3 R:12

Ad postN complete early July/late Aug.-early Sept./early Oct.; P duration variable: North Norway ♂ 48 days, ♀ 54 day, no difference between sexes in UK; P duration

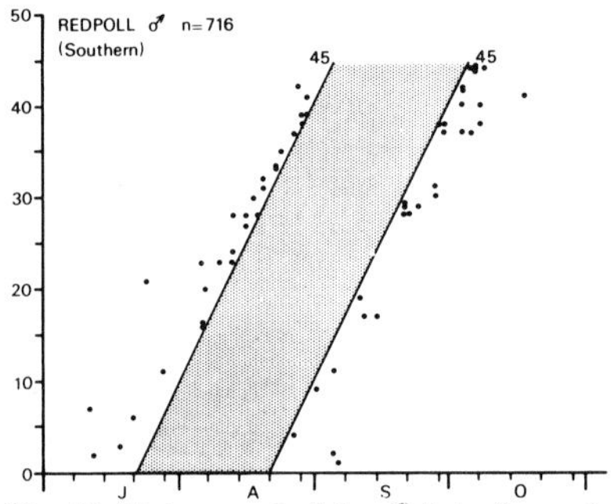

UKc 40-55 days, possibly slightly faster in North, c50 days in Finland; S start when Pscore 20-25 and finish same time as P; T start soon after P and full grown by Pscore 30; in captive birds P duration 44-58 days, S duration 32-45 days, R duration 27-33 days, B duration 39-58 days. **PJ** partial (B, lesser and median WC, some from early broods moult inner GC (not all), rarely 1-2 T) late July-Oct. **(74, 86, 137, 161, 162, 164, 234, 303, 323, 327, 405, 407, 610, 622, 623).**

CROSSBILL ***Loxia curvirostra*

P:9+1 m. S:6,E T:3 R:12

Most **Ad** have complete moult mid July-mid Nov., but due to irregular breeding Ad have been found in moult in all months, except Jan.-March; P duration at least 12 weeks. **PJ** partial (B, some inner GC), moult recorded in all months except Jan. In captive birds moult stated after breeding but in wild some breeding birds in moult; some Ad apparently suspend during 'irruptions'. Ad and Juv ♂ moulting before early July usually have yellowish feathers, while those moulting later usually have reddish ones, many birds which moult through this period have mixed plumage (change in colouration probably due to change in diet from old to new cones). In USA *benti* has partial **preN** moult in spring (head, throat). **(301,408,409,543,627, 628,630,665,694).**

BULLFINCH *****Pyrrhula pyrrhula*

P:9+1 m. S:6,E T:3 R:12

Ad postN complete mid July/early Sept.-late Sept./late Oct. P duration c60-75 days,

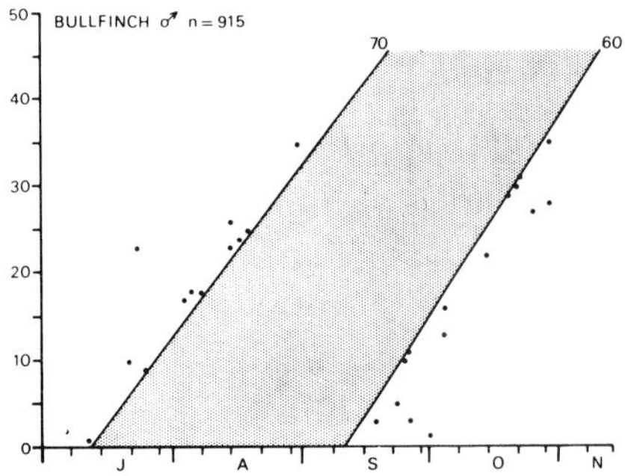

late starters moult more quickly; captive birds 84-89 days; in captive birds S duration 63-66 days finishing after P, in wild finish before P. **PJ** partial (B, T8) late Aug.-mid Oct.; a very few early moulting birds replace all GC (more in Europe?), most retain some old GC (usually 3, 4 or 5); old Carpal Covert usually retained, duration PJ moult 7-9 weeks. **(137,172,206,403,404,405,406,551, 608,623).**

HAWFINCH

***Coccothraustes coccothraustes*

P:9+1 m. S:6,E T:3 R:12

Ad postN complete mid July-end Sept., S start 2-3 weeks after P; UK Pscores: 30 July 17; 31 July 3,15; 2 Aug. 2; 6 Sept. 33; 13 Sept. 20; 5 Oct. 42,47,50,50. Moult duration c100 days in USSR. **PJ** partial (B, all GC?, very occasionally centre R and T) Aug.-Sept.**(135,249,351,387).**

EMBERIZIDAE

LAPLAND BUNTING
**Calcarius lapponicus*

P:9+1 m. S:6,E T:3 R:12

Ad postN complete July/Aug./Sept., timing of ♂ and ♀ similar; P duration 30-40 days, total moult c50 days; R simultaneous in USSR. **Ad preN** partial (head, upper throat) April. **PJ** partial (B, lesser and median WC) Aug.-Sept., starting c3 weeks after fledging. **(126,581,692,693, 710).**

SNOW BUNTING **Plectrophenax nivalis*

P:9+1 m. S:6,E T:3 R:12

Ad postN complete; mid July-late Aug. in NE Greenland, P start while still feeding fledged but dependent young, in Scotland B moult starts late July/early Aug. while feeding young; P duration c28 days NE Greenland, ♂ may start before ♀; S start when P1-P5 well grown and P6 and P7 dropped, finish by time P9 full grown, est duration 15 days; est T duration c17 days, T8 usually dropped 1st; R almost simultaneous, est duration c23 days. **Ad preN** 'moult' largely due to abrasion, but some moult on head March. **PJ** partial (B, lesser and median WC) Aug.-Sept., spring moult similar to Ad. **(14,205,298,401,434, 586,710).**

YELLOWHAMMER
****Emberiza citrinella*

P:9+1 m. S:6,E T:3 R:12

Ad postN complete early July/late Aug.-late Aug./late Oct. P duration c55 days; R

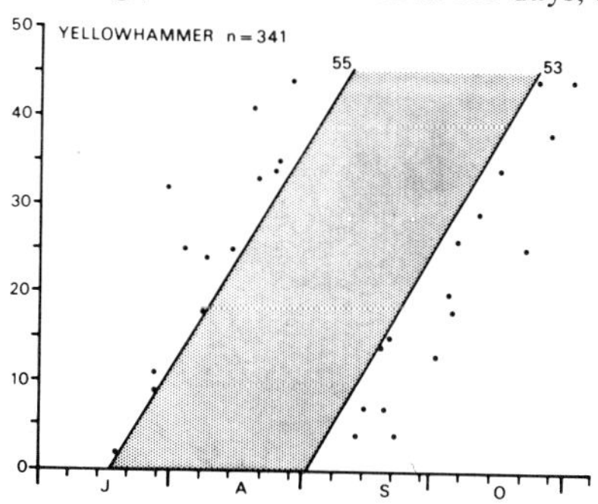

centrifugal. **PJ** partial (B, some (probably not all) GC, some T, sometimes some centre R) July-Oct. — no evidence of complete PJ moult (*c.f.* **146**). **(136,146,237,251, 298,405, 608,669).**

CIRL BUNTING **E. cirlus*

P:9+1 m. S:6,E T:3 R:12

No UK data. **Ad postN** complete, Aug.-Oct.; 49 in Sicily 16-30 Aug. av Pscore 29.5, range 8-39. **PreN** partial (head) March-April. **PJ** partial (B, all GC?, some T, sometimes some centre R) Aug.-Oct. **(710).**

REED BUNTING *****E. schoeniclus*

P:9+1 m. S:6,E T:3 R:12

Ad postN complete early July/late Aug.-early Aug./late Sept. P duration variable: c55 days Finland, 51-60 days Sweden, in UK early birds (starting c19 July) 50 days later birds (starting c7 Aug.) 33 days; the

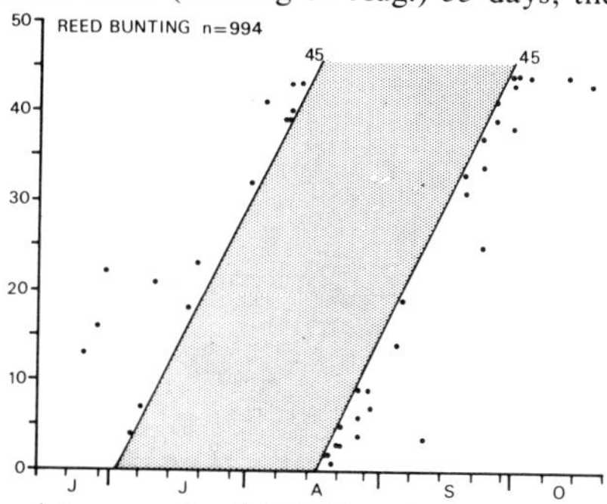

quicker moult of UK birds in comparison with Sweden and central Europe probably due to there being more breeding attempts per season in UK, moult starting after breeding; B starts with dropping P1 and continues after P finished; GC replaced centrifugally from centre of tract. **PreN** partial (head) combined with abrasion. **PJ** partial (B, inner GC (probably not all), probably all T), July-Oct.; often replaces centre R, early moulters more likely to replace all R than later birds; a few birds may occasionally undergo complete moult. **(57, 95, 166, 234, 237, 296, 297, 298, 327, 480, 547,622, 710).**

CORN BUNTING ****Miliaria calandra*

P:9+1 m. S:6,E T:3 R:12

Ad postN and **PJ** complete July-Oct. (occasionally later in Europe); P duration

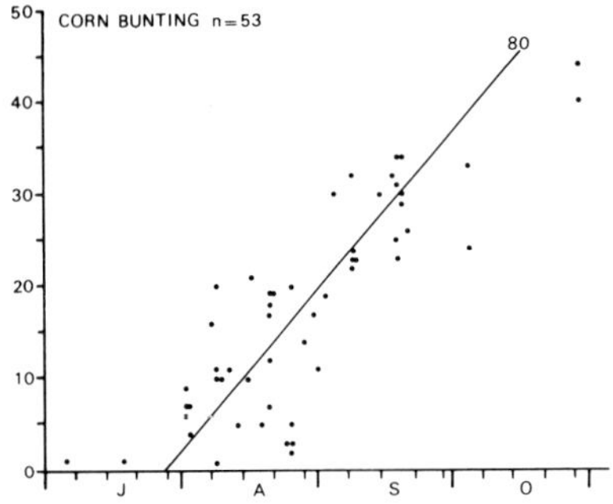

All birds, whether juvenile, adult or unaged, are included

c80 days, in some birds S and R not finished by Pscore 50. **(126,193,298,405).**

REFERENCES

a) References are listed in alphabetical sequence.
b) Reference numbers are those given in the species accounts in Part II.
c) Names starting with 'De' or 'Den' and 'Van' are listed under 'D' and 'V' respectively.
d) Titles of foreign papers are given in English if translated in the original paper. Square brackets [ ] indicate that there is no English summary in the original paper, round brackets () indicate that there is. Not all papers with an English summary translate the title, thus a number of papers with foreign language titles that certainly have an English summary are marked with an asterisk *.

ANON. 1972. **1**
Non-wader species caught, pp. 61-63 *in* PIENKOWSKI, M.W. (Ed.). *University of East Anglia Expedition to Morocco 1971 Report.* U.E.A. Expedition to Morocco, Norwich.

ANON. 1979. **2**
Primary moult. *Wash Wader Ringing Group Report* 1977-78: 62-63.

ANON. 1981. **3**
Wader primary moult sequences. *Wash Wader Ringing Group Report* 1979-80: 63-66.

AINLEY, D.G., LEWIS, T.J., and MORRELL, S. 1976 **4**
Moult in Leach's and Ashy Storm-petrels. *Wilson Bull.* 88: 76-95.

ALERSTAM, T. and HÖGSTEDT, G. 1982. **5**
Bird migration and reporduction in relation to habitats for survival and breeding. *Ornis Scand.* 13: 25-37.

AMADON, D. 1966. **6**
Avian plumages and moults. *Condor* 68: 263-278.

ANDERSEN-HARILD, P. 1981. **7**
Weight changes in *Cygnus olor. Proc. 2nd. Int. Swan Symp., Sapporo, Japan, 1980:* 359-378.

ANDERSON, B.W. and TIMKEN, R.L. 1971. **8**
Age and sex characteristics of Common Mergansers. *J. Wildl. Mgmt.* 35: 388-393.

ANDERSON, K.R. 1973. **9**
The biometrics, moult and recoveries of British-ringed Ruff. *Wader Study Group Bull.* 11: 5-14.

ANDRÉ, A. and FOURAGE J. 1967. **10**
Notes complémentaire concernant la mue du Pouillot fitis *(Phylloscopus trochilus). Aves* 4. 107-108.

ANKNEY, C.D. 1979. **11**
Does the wing moult cause nutritional stress in Lesser Snow Geese? *Auk* 69: 68-72.

APPLETON, G.F. and MINTON, C.D.T. 1978. **12**
The primary moult of the Lapwing. *Bird Study* 25: 253-256.

ASBIRK, S. 1979. **13**
A description of Danish Black Guillemots *Cepphus grylle* with remarks on the validity of *ssp. atlantis.* Dansk orn. Foren. Tidsskr. 73: 207-214.

ASBIRK, S. and FRANZMANN, N.-E. 1978. **14**
Studies of Snow Buntings, pp. 132-142 *in* GREEN, G.H. and GREENWOOD, J.J.D. (Eds.). *Joint biological expedition to North East Greenland 1974.* Dundee University N.E. Greenland Expedition, Dundee.

ASHMOLE, N.P. 1962. **15**
The Black Noddy *Anous tenuirostris* on Ascension Island. Part 1. General biology. *Ibis 103b: 235-273.*

ASHMOLE, N.P. 1963. **16**
The biology of the Wideawake or Sooty Tern *Sterna fuscata* on Ascension Island. Ibis 103b: 297-364.

ASHMOLE, N.P. 1968. **17**
Breeding and moult in the White Tern *(Gygis alba)* on Christmas Island, Pacific Ocean. *Condor* 70: 35-55.

ASHMOLE, N.P. 1971. **18**
Sea bird ecology and the marine environment. pp. 223-286 *in* FARNER, D.S. and KING, J.R. (Eds.). *Avian Biology* Vol. 1. Academic Press, London.

ASHMOLE, N.P., DORWARD, D.F., and **19**
STONEHOUSE, B. 1961. Numbering of primaries. *Ibis* 103a: 297-298.

ATKINSON, N.K., SUMMERS, R.W., NICOLL, M., **20**
and GREENWOOD, J.J.D. 1981. Population, movements and biometrics of the Purple Sandpiper *Calidris maritima* in eastern Scotland. *Ornis Scand.* 12: 18-27.

ATWOOD, J.L. and MASSEY, B.W. 1982. **21**
The *pikei* plumage of the Least Tern. *J. Field Ornithol.* 53: 47.

AVERILL, C.K. 1923. **22**
Black wing tips. *Condor* 25: 57-59.

AXTON, R.P. n.d. **23**
Studies of Long-tailed Tits. *Wicken Fen Group Report* 4 (1972): 32-33.

BAGGOTT, G.K. 1970. **24**
The timing of the moults of the Pied Wagtail. *Bird Study* 17: 45-46.

BAGGOTT, G.K. 1975. **25**
Moult, flight muscle "hypertrophy" and premigratory lipid deposition of the juvenile Willow Warbler, *Phylloscopus trochilus. J. Zool., Lond.* 175: 299-314.

BAGGOTT, G.K. n.d. **26**
The post-juvenile moult of the Pied Wagtail. *Chew Valley Ringing Station Report* 4 (1968-72): 18-21.

BAHR, P.H. 1909. **27**
On the supposed "colour-change" and the spring moult of the Black-headed Gull. *Br. Birds* 3: 105-111.

BÄHRMANN, U. 1958. **28**
Zur Mauser einiger Rabenvögel. *Vogelwelt* 79: 129-135.

BÄHRMANN, U. 1964. **29**
Über die Mauser des europäischen Stars *(Sturnus vulgaris* L.). *Zool. Abh. Staatl. Mus. Dresden* 27: 1-9.

BÄHRMANN, U. 1966. **30**
Die "Gesichtsmauser" der Saatkrähe *(Corvus frugilegus frugilegus* L.). *Zool. Abh. Mus. Tierk. Dresden* 28: 221-234.

BÄHRMANN, U. 1967. **31**
Bemerkungen zur Handschwingenmauser des Hausperlings *(Passer domesticus domesticus* L.). *Beitr. Vogelkunde* 12: 363-366.

BÄHRMANN, U. 1969. **32**
Einiges über die Flügel- und Schwanzmauser des Mäusebussards *(Buteo buteo buteo* L.). *Beitr. Vogelkunde* 14: 330-333.

BÄHRMANN, U. 1970. **33**
Über das Variieren der Schwanzmauser beim Star *(Sturnus vulgaris vulgaris* L.). *Beitr. Vogelkunde* 15: 434-436.

BÄHRMANN, U. 1971. **34**
Über eine regelwidridge Handschwingenmauser beim Eichelhäher *(Garrulus glandarius glandarius). Beitr. Vogelkunde.* 17: 413-414.

BAILLIE, S.R. and SWANN, R.L. 1980. **35**
The extent of post-juvenile moult in the Blackbird. *Ringing and Migration* 3: 21-26.

BAINBRIDGE, I. n.d.a. **36**
Biometrics and moult of Common Terns at Seaforth Dock, Liverpool. *Bird-Ringing in South West Lancashire* 7 (1977): 34-40.

BAINBRIDGE, I.P. n.d.b. **37**
Plumage and biometrics of Little Gulls on passage in South West Lancashire. *Bird Ringing in South West Lancashire* 8 (1978): 49-52.

BAINBRIDGE, I.P. and BAINBRIDGE, C.A. n.d. **38**
Plumage, biometrics and moult of Kittiwakes in South-West Lancashire. *Bird Ringing in South West Lancashire* 8 (1978): 32-35.

BAINBRIDGE, I.P. and MINTON, C.D.T. 1979. **39**
Primary moult of Curlews at the Wash, England. *Wader Study Group Bull.* 29: 28-29.

BAKER, K. 1979. **40**
A guide to ageing and sexing non-passerines. Part 1. *Ringers' Bulletin* 5: 82-86.

BAKER, K. 1980a. **41**
Guide to ageing and sexing non-passerines. Part 2. *Ringers' Bulletin* 5: 91-94.

BAKER, K. 1980b. **42**
Guide to ageing and sexing non-passerines. Part 3. *Ringers' Bulletin* 5: 104-107.

BAKER, K. 1981a. **43**
Guide to ageing and sexing non-passerines. Part 4. *Ringers' Bulletin* 5: 117-119.

BAKER, K. 1981b. **44**
Guide to ageing and sexing non-passerines. Part 5. *Ringers' Bulletin* 5: 133-135.

BALAT, F. 1960. **45**
(A study of moulting in the Dipper, *Cinclus cinclus* (L.).). *Zool. Listy* 9: 257-264.

BALAT, F. 1970. **46**
On the wing-moult in the Mallard *Anas platyrhynchos* in Czechoslovakia. *Zool. Listy* 19: 135-144.

BALLANTYNE, J.H. and DA PRATO, S.R.D. n.d. **47**
Sedge Warbler, Willow Warbler and Whitethroat at Cousland 1973-76. *Edinburgh Ringing Group Report* 4 (1976): 16-30.

BANCROFT, G.T. and WOOLFENDEN, G.E. 1982. **48**
The moult of Scrub Jays and Blue Jays in Florida. *A.O.U. Ornithological Monographs* No. 29.

BANKS, R.C. 1978. **49**
Prealternate molt in Nuthatches. *Auk* 95: 179-181.

BARTH, E.K. 1975. **50**
Moult and taxonomy of the Herring Gull *Larus argentatus* and the Lesser Black-backed Gull *L. fuscus* in northwestern Europe. *Ibis* 117: 384-387.

BAUER, K.M. and GLUTZ VON BLOTZHEIM, U.N. **51**
1966. *Handbuch der Vögel Mitteleuropas.* Vol. 1. Akademische Verlagsgesellschaft, Frankfurt am Main.

BAYNE, C.S. 1918. **52**
Notes on the moulting of the Black-headed Gull. Paper read before the Ornithological Section of the London Natural History Society. MS.

BECKER, P. 1974. **53**
Beobachtungen an paläarktischen Zugvögeln in ihrem Winterquartier in Südwestafrika. *Wiss. Forschungin Südwestafrika,* 12. Folge.

BECKER, P. 1978. **54**
Sommerliche Schlaf- und Mausergemeinschaft der Waldohreule *(Asio otus). Vogelkundliche Ber Niedersachs.* 10: 23-24.

BECKER, P. and LÜTGENS, H. 1976. **55**
Sumpfrohrsänger *(Acrocephalus palustris)* in Südwestafrika. *Madoqua* 9: 41-44.

BEINTEMA, A.J. and MÜSKENS, G.J.D.M. 1981. **56**
Veranderingen in de trekgewoonten van de watersnip *(Gallinago gallinago)* in Europa en de invloed van de mens hierop. RIN rapport 81/1. Rijksinstituut voor Natuurbeheer, Leersum.

BELL, B.D. 1970. **57**
Moult in the Reed Bunting — a preliminary analysis. *Bird Study* 17: 269-281.

BENGTSON, S. -A. 1975. **58**
Timing of the moult of the Purple Sandpiper *Calidris maritima* in Spitsbergen. *Ibis* 117: 100-101.

BERGER, W. 1967. **59**
Die Mauser des Sprossers *(Luscinia luscinia* L.). *J. Orn.* 108: 320-327.

BERTHOLD, P. 1973. **60**
Circannuale Periodik bei Teilziehern und Standvögeln. *Naturwissenschaften* 60: 252-253.

BERTHOLD, P. 1974. **61**
(Circannual rhythms in warblers *(Sylvia).* III. Rhythms of moult, nocturnal restlessness, and of body weight changes in Mediterranean species with different migratory habits). *J. Orn.* 115: 251-272.

BERTHOLD, P., GWINNER, E., and KLEIN, H. 1970. **62**
(Comparative study of the juvenile development of a typical migrant, the Garden Warbler, and of a less typical migrant, the Blackcap). *Vogelwarte* 25: 297-331.

BERTHOLD, P., GWINNER, E., and KLEIN, H. 1972. **63**
(Circannual rhythms in warblers. I. Rhythms of body weight, molt, and nocturnal restlessness in *Sylvia atricapilla* and *S. borin* under different constant conditions). *J. Orn.* 113: 170-189.

BIANKI, V.V. 1977. **64**
Gulls, shorebirds and alcids of Kandalaksha Bay. Israel Program for Scientific Translations, Jerusalem.

BIBBY, C.J. 1977. **65**
Observations on the moult of the Tree Sparrow. *Ringing and Migration* 1: 148-157.

BIBBY, C.J. n.d. **66**
Post juvenile moult of the Tree Sparrow. *Wicken Fen Group Report* 2 (1970): 10-20.

BILLARD, R.S. and HUMPHREY, P.S. 1972 **67**
Molts and plumages of the Greater Scaup. *J. Wildl. Mgmt.* 765-774.

BIRKHEAD, T.R. and TAYLOR, A.M. 1977. **68**
Moult of the Guillemot *Uria aalge. Ibis* 119: 80-85.

BLACKMORE, F.H. 1969. **69**
The effect of temperature, photoperiod and molt on the energy requirements of the House Sparrow, *Passer domesticus. Comp. Biochem. Physiol.* 30: 433-444.

BLÜMEL, H. 1976. **70**
Der Grünling. Neue Brehm-Bücherei, Wittenberg Lutherstadt.

BOASE, H. 1959. **71**
Notes on the display, nesting and moult of the Mute Swan. *Br. Birds* 52: 114-123.

BODDY, M. 1981. **72**
Primary moult in juvenile and adult Woodpigeons. *Ringing and Migration* 3: 113-122.

BODDY, M. in press. **73**
Factors influencing timing of autumn dispersal or migration in first-year Dunnocks and Whitethroats.

BODDY, M. n.d. **74**
Post-nuptial and post-juvenile moults of the Lesser Redpoll *(Carduelis flammea cabaret)* in Nottinghamshire, England, and their relationships to breeding and to migration. MS.

BOERE, G.C. 1974. **75**
Wader research in the Dutch part of the Waddensea. *Proc. I.W.R.B. Wader Symposium, Warsaw, 1973:* 83-98.

BOERE, G.C. 1976. **76**
The significance of the Dutch Waddenzee in the annual life cycle of arctic, subarctic and boreal waders. Part 1. The function as a moulting area. *Ardea* 64: 210-291.

BOERE, G.C., BRUIJNE, J.W.A. de, and NIEBOER, E. **77**
1973. (A study on the significance of the Dutch Waddensea-area for the Dunlin *Calidris alpina* in late summer and autumn). *Limosa* 46: 205-227.

BORRETT, R.P. 1971. **78**
Movements and moult in the Garden Warbler in Rhodesia. *Ostrich* 42: 141-142.

BRANSON, N.J.B.A. and MINTON, C.D.T. 1976. **79**
Moult, measurements and migrations of the Grey Plover. *Bird Study* 23: 257-266.

BRANSON, N.J.B.A., PONTING, E.D. and **80**
MINTON, C.D.T. 1979. Turnstone populations on the Wash. *Bird Study* 26: 47-54.

BRITTON, P.L. 1978. **81**
Seasonality, density and diversity of birds of a papyrus swamp in western Kenya. *Ibis* 120: 450-466.

BROEKHUYSEN, G.J. and BROWN, A.R. 1963. **82**
The moulting pattern of European Swallows, *Hirundo rustica*, wintering in the surroundings of Cape Town, South Africa. *Ardea* 51: 25-43.

BROEKHUYSEN, G.J. and SIEGFRIED, W.R. 1970. **83**
Age and moult in the Steppe Buzzard in Southern Africa. *Ostrich Suppl.* 8: 223-237.

BROM, T.G. 1980. **84**
Microscopic identification of feather-remains after collisions between birds and aircraft. Instituut voor Taxonomische Zoologie, Amsterdam. 89pp.

BROOKE, R.K. 1974. **85**
The Spotted Crake *Porzana porzana* (Aves: Rallidae) in south-central and southern Africa. *Durban Mus. Novit.* 10: 43-52.

BROOKS, W.S. 1968. **86**
Comparative adaptations of the Alaskan Redpolls to the arctic environment. *Wilson Bull.* 80: 253-280.

BROWN, L. and AMADON, D. 1968. **87**
Eagles, hawks and falcons of the World. Country Life, London.

BROWN, S.C. 1973. **88**
Common Sandpiper biometrics. *Wader Study Group Bull.* 11: 18-24.

BRYANT, D.M. 1978. **89**
Moulting Shelducks on the Forth Estuary. *Bird Study* 25: 103-108.

BRYANT, D.M. 1981. **90**
Moulting Shelducks on the Wash. *Bird Study* 28: 157-158.

BUKER, J.B., BUURMA, L.S. and OSIECK, E.R. **91**
1975. Post-juvenile moult of the Bearded Tit, *Panurus biarmicus* (Linnaeus, 1758), in Zuidelifk Flevoland, the Netherlands (Aves, Timaliinae). *Beaufortia* 23: 169-179.

BURGER, J.W. 1941. **92**
Experimental modification of the plumage cycle of the male European Starling. *Bird-Banding* 12: 27-29.

BURI, A. 1967. **93**
Das Juvenilgefieder von *Phasianus colchicus* L., ein Beitrag zur Kenntnis dieser Altersetappe des Gefieders. *Revue suisse Zool.* 74: 301-387.

CARRICK, R. and DUNNET, G.M. 1954. **94**
Breeding of the Fulmar *Fulmarus glacialis*. *Ibis* 96: 356-370.

CARTHY, D.P. 1973. **95**
Abnormal Reed Bunting moult. *Proceedings Coventry Distr. nat. Hist. scient. Soc.* 4: 132-133.

CASTO, S.D. 1974. **96**
Molt schedule of House Sparrows in northwestern Texas. *Wilson Bull.* 86: 176-177.

CHANDLER, A.C. 1916. **97**
A study of the structure of feathers, with reference to their taxonomic significance. *Univ. Calif. Publs Zool.* 13: 243-446.

CHAPIN, J.P. 1921. **98**
The abbreviated inner primaries of nestling woodpeckers. *Auk* 38: 531-552.

CLANCEY, P.A. 1948. **99**
On the partial summer moult of *Cuculus canorus* Linnaeus. *Ibis* 90: 333.

CLAPHAM, C. 1978. **100**
The Ringed Plover populations of Morecambe Bay. *Bird Study* 25: 175-180.

CLAPHAM, C. 1979. **101**
The Turnstone populations of Morecambe Bay. *Ringing and Migration* 2: 144-150.

CLARK N.A. 1978 **102**
The weights, moult and morphometrics of Spotted Redshanks in Britain. *Wader Study Group Bull.* 22: 22-26.

CLARK, P.N.J. n.d. **103**
Rate of moult in Tree Sparrows. *Tay Ringing Group Report* 1975: 11-17.

CLAUSAGER, I. 1973. **104**
Age and sex determination of the Woodcock *(Scolopax rusticola). Danish Review of Game Biology* 8: 1-18.

CLENCH, M.H. 1970. **105**
Variability in body pterylosis, with special reference to the genus Passer. *Auk* 87: 650-691.

CLENCH, M.H. 1973. **106**
Variability in the body pterylosis of *Passer domesticus* and allies. pp. 32-33 *in* KENDEIGH, S.C. (Chairman). A symposium on the House Sparrow *(Passer domesticus)* and European Tree Sparrow *(P. montanus)* in North America. *A.O.U. Ornithological Monographs* No. 14.

COHEN, J. 1966. **107**
Feathers and patterns. *Advan. Morphog.* 5: 1-38.

COLQUHOUN, M.K. 1951. **108**
The Wood Pigeon in Britain. A.R.C. Report Series 10. H.M.S.O., London.

COMPTON, L.V. 1938. **109**
The pterylosis of the Falconiformes with special reference to the taxonomic position of the Osprey. *Univ. Calif. Publs Zool.* 42: 173-212.

COOMBS, C.F.B., ISAACSON, A.J., MURTON, **110**
R.K., THEARLE, R.J.P. and WESTWOOD, N.J. 1981. Collard Doves *(Streptopelia decaocto)* in urban habitats. *J. appl. Ecol.* 18: 41-62.

CRAMP, S. and SIMMONS, K.E.L. (Eds.). 1977, **111**
1980. *in press. Handbook of the birds of Europe, the Middle East, and North Africa: the birds of the Western Palearctic.* Vols 1,2,3. Oxford University Press, Oxford.

CREUTZ, G. 1955. **112**
Der Trauerschnäpper *(Muscicapa hypoleuca* (Pallas)). Eine Populations-studie. *J. Orn.* 96: 241-326.

CROXALL, J.P., 1982.
Energy costs of incubation and moult in petrels and penguins. *J. Anim. Ecol.* 51: 177-194.

CULLEN, J.M. 1957. **113**
Plumage, age and mortality in the Arctic Tern. *Bird Study* 4: 197-207.

CURRY-LINDAHL, K. 1963. **114**
Molt, body weights, gonadal development, and migration in *Motacilla flava. Proc. XIII Intern. Ornithol. Congr.* 960-973.

DAANJE, A. 1935. **115**
Waarnemingen over den vroegen zomertrek en den slagpenrui van den Kievit *(Vanellus vanellus* (L.)) in de jaren 1933 en 1934. *Ardea* 24: 86-97.

DARE, P. J. and MERCER, A.J. 1974. **116**
The timing of wing-moult in the Oystercatcher *Haematopus ostralegus* in Wales. *Ibis* 116: 211-214.

DAVIES, S.J., 1981. **117**
Skylarks at Gibraltar Point, Lincolnshire. *Ringing and Migration* 3: 173-180.

DAY, M.G. 1966. **118**
Indentification of hair and feather remains in the gut and faeces of stoats and weasels. *J. Zool., Lond.* 148: 201-217.

DEAN, W.R.J. 1977a. **119**
Moult of Little Stints in South Africa. *Ardea* 65: 73-79.

DEAN, W.R.J. 1977b. **120**
Moult of the Curlew Sandpiper at Barberspan. *Ostrich Suppl.* 12: 97-101.

DE BONT, A.F. 1962. **121**
Composition des bandes d'Hirondelles de cheminée *Hirundo rustica rustica* L. hivernant au Katanga et analyse de la mue des rémiges primaires. *Gerfaut* 52: 298-343.

DE BONT, A.F. 1970 **122**
Some results of bird-banding in the Congo (Kinshasa). *Ostrich* 41: 195-199.

DECKERT, G. 1962. **123**
Zur Ethologie des Feldsperlings *(Passer m. montanus L.). J.*

Orn. 103: 428-486.

DEHAEN, M and HERROELEN, P. 1971. **124**
Enkele gevallen van onderbroken rui bij de Tuinfluiter *(Sylvia borin). Gerfaut* 61: 104-106.

DEKKER, D. 1963. **125**
Serie-onderzoek bij Turkse Tortels *(Streptopelia decaocto). Limosa* 36: 141-148.

DEMENT'EV, G.P. and GLADKOV, N.A. (Eds.). **126**
1966-1970. *Birds of the Soviet Union.* Vols I-VI. Israel Program for Scientific Translations, Jerusalem.

DEN BLANKEN, G., BOERE, G.C. and NIEBOER, E. **127**
1981. Primary moult of the Redshank *Tringa totanus* in the Dutch Waddenzee studied by collecting shed feathers: a test. *Ardea* 69: 115-124.

DE ROO, A. 1966. **128**
Age-characteristics in adult and sub adult Swifts, *Apus a. apus* (L.), based on interrupted and delayed wing-moult. *Gerfaut* 56: 113-131.

DHONDT, A.A. 1973. **129**
Postjuvenile and postnuptial moult in a Belgian population of Great Tits, *Parus major,* with some data on captive birds. *Gerfaut* 63: 187-209.

DHONDT, A.A. 1981. **130**
Postnuptial moult of the Great Tit in southern Sweden. *Ornis Scand.* 12: 127-133.

DIAMOND, A.W. 1974. **131**
Annual cycles in Jamaican forest birds. *J. Zool., Lond.* 173: 277-301.

DIAMOND, A.W. 1976. **132**
Subannual breeding and moult cycles in the Bridled Tern *Sterna anaethetus* in the Seychelles. *Ibis* 118: 414-419.

DICK, W.J.A. 1975. **133**
Wader research at the Banc d'Arguin. pp. 12-46 *in* DICK, W.J.A. (Ed.). *Oxford and Cambridge Mauritanian Expedition 1973 Report.* Cambridge.

DIESSELHORST, G. 1961. **134**
Ascendente Handschwingen-Mauser bei *Muscicapa striata. J. Orn.* 102: 360-366.

DOLNIK, V.R. 1965. **135**
[Bioenergetics of the moults of finches as adaptations to migration]. *Novosti Ornithol.* 4: 124-126.

DOLNIK, V.R. and BLYUMENTAL, T.I. 1967. **136**
Autumnal premigratory and migratory periods in the Chaffinch *(Fringilla coelebs coelebs)* and some other temperate-zone passerine birds. *Condor* 69: 435-468.

DOLNIK, V.R. and GAVRILOV, V.M. 1974. **137**
[Half-quantitative method of registration of moult in passerine birds]. *Ornithologia 11: 110-125.*

DOLNIK, V.R. and GAVRILOV, V.M. 1979. **138**
Bioenergetics of molt in the Chaffinch *(Fringilla coelebs). Auk* 96: 253-264.

DOLNIK, V.R. and GAVRILOV, V.M. 1980. **139**
Photoperiodic control of the molt cycle in the Chaffinch *(Fringilla coelebs). Auk* 97: 50-62.

DONNER, E. 1907. **140**
Die Mauser des Baumfalken *(Falco subbuteo). J. Orn.* 55: 583-591.

DORWARD, D.F. 1963. **141**
The Fairy Tern *Gygis alba* on Ascension Island. *Ibis* 103b: 365-378.

DOUCET, J. 1971. **142**
(The moult of remiges and rectrices in the Kingfisher *(Alcedo atthis)). Gerfaut* 61: 14-42.

DOWSETT, R.J. 1966. **143**
The moulting pattern of European Swallows, *Hirundo rustica,* wintering in Eastern Zambia. *Puku* 4: 91-100.

DOWSETT, R.J. 1971. **144**
Suspended wing moult in migrants. *Bird Study* 18: 53-54.

DRENCKHAHN, D. 1968. **145**
Die Mauser des Kampfläufers, *Philomachus pugnax,* in Schleswig-Holstein. *Corax* 2: 130-150.

DROST, R. 1969. **146**
Grundsätzliches zur Altersbestimmung lebender Sperlingsvögel. *Vogelwarte* 25: 6-13.

DUBBELDAM, W. 1978. **147**
(The Greylag Goose *Anser anser* in Flevoland in 1972-1975). *Limosa* 51: 6-30.

DUNNET, G.M. 1955. **148**
The breeding biology of the Starling *Sturnus vulgaris* in relation to its food supply. *Ibis* 97: 619-662.

DUNNET, G.M., FORDHAM, R.A. and **149**
PATTERSON, I.J. 1969. Ecological studies of the Rook *(Corvus frugilegus* L.) in North-East Scotland. Proportion and distribution of young in the population. *J. appl. Ecol.* 6: 459-473.

DWIGHT, J. 1900. **150**
The sequence of plumages and moults of the passerine birds of New York. *Ann. New York Acad. Sci.* 13: 73-360.

DWIGHT, J. 1901. **151**
The sequence of moults and plumages of the Laridae (gulls and terns). *Auk* 18: 49-63.

DWIGHT, J. 1925. **152**
The gulls (Laridae) of the world; their plumages, moults, variations, relationships and distribution. 52: 63-336.

ELLIOT, C.C.H., WALTNER, M., UNDERHILL, **153**
L.G., PRINGLE, J.S. and DICK W.J.A. 1976. The migration system of the Curlew Sandpiper *Calidris ferruginea* in Africa. *Ostrich* 47: 191-213.

ELTRINGHAM, S.K. and BOYD, H. 1960. **154**
The Shelduck population in the Bridgwater Bay moulting area. *Wilfowl Trust Ann. Rep.* : 10

ENDERSON, J.H., TEMPLE, S.A. and **155**
SWARTZ, L.G. 1972. Time-lapse photographic records of nesting peregine falcons. *Living Bird* 11: 113-128.

ERKINARO, E. 1975. **156**
Zeitpunkt und Dauer der Mauser des Rauhfusskauzes, und der Sumpfohreule, *Asio flammeus.*
Beitr. Vogelkund 21: 288-290.

ERSKINE, A.J. 1971. **157**
Growth, and annual cycles in weights, plumages and reproductive organs of Goosanders in Eastern Canada. *Ibis* 113: 42-58.

ESCALENTE, R. 1970. **158**
Notes on the Cayenne tern in Uruguay. *Condor : 89-94.*

'ESPINASSE, P.G. 1964. **159**
Feather. pp. 272-277 *in* THOMSON, A.L. (Ed.). *A new dictionary of birds.* Nelson, London.

ETHERIDGE, B. 1980. **160**
Wader studies. *in* POMEROY, C.A. (Ed.). *Report on the RAFOS Masirah Island Expedition 22nd October-26th November 1979.*

EVANS, P.R. 1966. **161**
Autumn movements, moult and measurements of the Lesser Redpoll *Carduelis flammea cabaret. Ibis* 108: 183-216.

EVANS, P.R. 1969. **162**
Ecological aspects of migration, and pre-migratory fat deposition in the Lesser Redpoll, *Carduelis flammea cabaret. Condor* 71: 316-330.

EVANS, P.R. 1971. **163**
Moulting seasons of Willow Warblers and Bramblings in Norwegian Lapland. *Ornis Fenn.* 48: 131-132.

EVANS, P.R., ELTON, R.A. and SINCLAIR, G.R. **164**
1967. Moult and weight changes of Redpolls, *Carduelis flammea,* in north Norway. *Ornis Fenn.* 44: 33-41.

FAGAN, M.J., SCHMITT, M.B. and **165**
WHITEHOUSE, P.J. 1976. Moult of Purple Gallinule and Moorhen in the southern Transvaal. *Ostrich* 47: 226-227.

FENNELL, J.F.M. and STONE, D.A. 1976. **166**
A winter roosting population of Reed Buntings in central England. *Ringing and Migration* 1: 108-114.

FERNS, P.N. 1978. **167**
The onset of prebasic body moult during the breeding season in some high-Arctic waders. *Bull. Br. Orn. Club* 98: 118-122.

FERNS, P.N. and GREEN, G.H. 1979. **168**
Observations on the breeding plumage and prenuptial moult of Dunlins, *Calidris alpina,* captured in Britain. *Gerfaut* 69: 286-303.

FJELDSÅ J. 1973. **169**
Distribution and geographical variation of the Horned Grebe *Podiceps auritus* (Linnaeus, 1758). *Ornis Scand.* 4: 55-86.

FLEGG, J.J.M. 1973. **170**
A study of Treecreepers. *Bird Study* 20: 287-302.

FLEGG, J.J.M. and COX, C.J. 1969. **171**
The moult of British Blue Tit and Great Tit populations. *Bird Study* 16: 147-157.

FLEGG, J.J.M. and MATTHEWS, N.J. 1980. **172**
Ageing Bullfinches. *Ringers' Bulletin* 5: 95.

FLINT, P.R. 1972. **173**
Moult in Cetti's Warbler *Cettia cetti*. *Cyprus Ornithological Society 2nd Bird Report* 1971: 109.

FOG, J. 1969. **174**
(Studies on the Coot *(Fulica atra)* in the Marshland Vejlerne, North Jutland, and on the migration of Danish breeding populations). *Dansk orn. Foren. Tidsskr.* 63: 1-18.

FOGDEN, M.P.L. 1972. **175**
The seasonality and population dynamics of equatorial forest birds in Sarawak. *Ibis* 114: 307-343.

FORD, H.A. 1980. **176**
Breeding and molt in Honeyeaters (Aves: Meliphagidae) near Adelaide, South Australia. *Aust. Wildl. Res.* 7: 453-463.

FOSTER, M.S. 1975. **177**
The overlap of molting and breeding in some tropical birds. *Condor* 77: 304-314.

FOURAGE, J. 1967. **178**
A propos de la mue post-nuptiale du Pouillot fitis *(Phylloscopus trochilus)*. *Aves* 4: 69.

FRANCIS, D.M. 1980. **179**
Moult of European Swallows in Central Zambia. *Ringing and Migration* 3: 4-8.

FRELIN, C. 1977. **180**
Analyse biométrique des captures de mésanges bleues *Parus caeruleus* au col de la Golèze (Haute-Savoie). *Alauda* 45: 105-113.

FUCHS, E. 1973. **181**
Durchzug und Überwinterung des Alpenstrandläufers *Calidris alpina* in der Camargue. *Orn. Beob.* 70: 113-134.

FURNESS, R.W. and BAILLIE, S.R. 1981. **182**
Age ratios, wing length and moult as indicators of the population structure of Redshank wintering on British estuaries. *Ringing and Migration* 3: 123-132.

GALBRAITH, H. 1977. **183**
The post-nuptial moult of a migratory population of Pied Wagtails. *Ringing and Migration* 1: 184-186.

GALBRAITH, H., MITCHELL, A.B. and **184**
SHAW, G. 1981. The moult of the Dipper in central Scotland. *Bird Study* 28: 53-59.

GASTON, A.J. 1981. **185**
Seasonal breeding, moulting and weight changes among birds of dry deciduous forest in North India. *J. Zool., Lond.* 194: 219-243.

GAUCKLER, A and KRAUS, M. 1968. **186**
Zum Vorkommen und zur Brutbiologie des Schwarzhalstauchers *(Podiceps nigricollis)* in Nordbayern. *Anz. orn. Ges. Bayern* 8: 349-364.

GAVRILOV, E. 1971. **187**
(On moult of primaries in swallows during the autumn flight). *Zool. Zh.* 50: 599-602.

GAVRILOV, V.M. 1979. **188**
[Moult characteristics of sedentary and migrative subspecies of the Chaffinch *(Fringilla coelebs)* and House-Sparrow *(Passer domesticus)*]. *Ornithologiya* 14: 158-163.

GINN, H.B., 1973. **189**
Moult recording. *Auspicium* 5 (Suppl): 39-48.

GINN, H.B. 1975. **190**
The timing and sequence of the complete annual moult in the Dunnock *(Prunella modularis)* in Britain over an eleven year period. *J. Orn.* 116: 263-280.

GINN, H. and GLUE, D. 1974. **191**
Transilient and irregular moult in the Tawny Owl. *Bird Study* 21: 87-88.

GLADWIN, T.W. 1969. **192**
Post-nuptial wing-moult in the Garden Warbler. *Bird Study* 16: 131-132.

GLIEMANN, L. 1973. **193**
Die Grauammer. Neue Brehm-Bücherei, Wittenberg Lutherstadt.

GLUTZ VON BLOTZHEIM, U.N. 1972. **194**
(On the moult of *Charadrius hiaticula, dubius* and *alexandrinus*). *J. Orn.* 113: 323-333.

GLUTZ VON BLOTZHEIM, U.N., BAUER, K.M. **195**
and BEZZEL, E. 1975. *Handbuch der Vögel Mitteleuropas.* Vol. 6. Akademische Verlagsgesellschaft, Weisbaden.

GLUTZ VON BLOTZHEIM, U.N. and **196**
BAUER, K.M. 1980. *Handbuch der Vögel Mitteleuropas.* Vol. 9. Akademische Verlagsgesellschaft, Wiesbaden.

GOETHE, F. 1961. **197**
A survey of moulting Shelduck on Knechtsand. *Br. Birds* 54: 106-115.

GOODYER, L.R. 1976. **198**
Lapwing weights and moult. *Wader Study Group Bull.* 18: 9-12.

GRANT, C.H.B. 1914. **199**
The moults and plumages of the Common Moorhen *(Gallinula chloropus* Linn.). *Ibis.* (Ser. 10) 2: 298-304.

GRANT, P.J. 1982. **200**
Gulls. A guide to identification. Poyser, Calton.

GRANT, P.J., SCOTT, R.E. and **201**
WALLACE, D.I.M. 1971. Further notes on the 'portlandica' plumage phase of terns. *Br. Birds* 64: 19-22.

GRAUBITS, G. and JANAUS, M. 1978. **202**
(Moult migration of the Common Scoter on the Kolkasrags Spit in July and August, 1975). *Soobshcheniya pribalt. Kom. Izuch. Migr. Ptits* No. 11: 37-45.

GREEN, G.H. 1973. **203**
Some notes on Bar-tailed Godwit ringing, biometrics and moult. *Wader Study Group Bull.* 8: 4-8.

GREEN, G.H. 1978. **204**
Wintering gulls in Worcestershire and Warwickshire. Progress Report No. 2. (duplicated).

GREEN, G.H. and SUMMERS, R.W. 1975. **205**
Snow Bunting moult in northeast Greenland. *Bird Study* 7.

GREEN, R.E. n.d. **206**
A new approach to moult studies. *Wicken Fen Group Report* 6 (1974): 37-41.

GURR, L. 1954. **207**
A study of the Blackbird *Turdus merula* in New Zealand. *Ibis* 96: 225-261.

GWINNER, E. 1966. **208**
Der zeitliche Ablauf der Handschwingenmauser des Kolkraben *(Corvus corax* L.) und seine funktionelle Bedeutung. *Vogelwelt* 87: 129-133.*

GWINNER, E. 1968. **209**
Circannuale Periodik als Grundlage des jahreszeitlichen Funktionswandels bei Zugvögeln. *J. Orn.* 109: 70-95.

GWINNER, E. 1969. **210**
Untersuchungen zur Jahresperiodik von Laubsängern. *J. Orn.* 110: 1-21.

GWINNER, E.G. 1972. **211**
Adaptive functions of circannual rhythms in warblers. *Proc. XVth Int. Orn. Congr., The Hague, 1970:* 218-236.

GWINNER, E. 1973. **212**
Die Dauer der Grossfiedermauser beim Fitis *(Phylloscopus trochilus)* und beim Zilpzalp *(Ph. collybita)* unter verschiedenen photoperiodischen Bedingungen. *J. Orn.* 114: 507-510.

GWINNER, E. 1977. **213**
Photoperiodic synchronization of circannual rhythms in the European Starling *(Sturnus vulgaris)*. *Naturwissenschaften* 64: 44.

GWINNER, E., BERTHOLD, P. and **214**
KLEIN, H. 1972. (Investigations on annual rhythms in Warblers. III. The development of plumage, body weight and migratory restlessness in Scandinavian and southwest German willow warblers *(Phylloscopus trochilus acredula* and *Ph. t.*

trochilus) under identical conditions). *J. Orn.* 113: 1-8.

GWINNER, E. and BIEBACH, H. 1977. **215**
(Endogenous control of moult and migratory disposition in red-backed shrikes *(Lanius collurio)* from a Finnish and a French population). *Vogelwarte* 29: 56-63.

HALD-MORTENSEN, P. 1970. **216**
(Fulmars *(Fulmarus glacialis)* in Skager-Rak). *Dansk orn. Foren. Tidsskr.* 64: 273-274.

HALE, J.W. n.d.a. **217**
Study of Long-tailed Tit moult in Gloucestershire 1976. *Severn Vale Ringing Group Report* 11 (1976): 29-31.

HALE, J.W. n.d.b. **218**
Long-tailed Tit moult in 1977 compared with the 1976 moult. *Severn Vale Ringing Group Report* 12 (1977): 48-50.

HALE, W.G. 1971. **219**
A revision of the taxonomy of the Redshank *Tringa totanus*. *Zool. J. Linn. Soc.* 50: 199-268.

HALE, W.G. 1980. **220**
Waders. Collins, London.

HALL, B.P. 1961. **221**
The taxonomy and identification of pipits (genus *Anthus*). *Bull. Br. Mus. nat. Hist. Zoology* 7: 243-289.

HANMER, D.B. 1979. **222**
A trapping study of Palaearctic passerines at Nchalo, Southern Malawi. *Scopus* 3: 81-92.

HANMER, D.B. 1981. **223**
Abnormal numbers of rectrices. *Safring News* 10: 3-5.

HARDY, A.R., HIRONS, G.J.M. and STANLEY, P.I. 1981. The relationship of body weight, fat deposit and moult to the reproductive cycles in wild Tawny Owls and Barn Owls. pp. 159-163 *in* COOPER, J.E. and GREENWOOD, A.G. (Eds.). *Recent advances in the study of raptor diseases*. Chiron Publications, Keighly. **224**

HARRIS, M.P. 1971. **225**
Ecological adaptations of moult in some British gulls. *Bird Study* 18: 113-118.

HARRIS, M.P. 1981. **226**
Age determination and first breeding of British Puffins. *Br. Birds* 74: 246-256.

HARRIS, M.P. and YULE, R.F. 1977. **227**
The moult of the Puffin *Fratercula arctica*. *Ibis* 119: 535-541.

HARRISON, C.J.O. 1963. **228**
'Industrial' discoloration of House Sparrows and other birds. *Br. Birds* 56: 296-297.

HARRISON, M.C. 1975. **229**
Tern ringing studies. pp. 47-51 *in* DICK, W.J.A. (Ed.). *Oxford and Cambridge Mauritanian Expedition 1973 Report*. Cambridge.

HARTERT, E. 1910-1922. **230**
Die Vögel der Paläarktischen Fauna. Friedländer, Berlin.

HARTMANN-MÜLLER, B. 1973. **231**
Beiträge zur Mauser des Steinkauzes, *Athene noctua* (Scop.). 1. Der Handflügel. *Philippia, Abh. Ber. Naturkde Mus. Kassel* 1: 286-295.*

HARTMANN-MÜLLER, B. 1974. **232**
Beiträge zur Mauser des Steinkauzes, *Athene noctua* (Scop.). 2. Armflugel und Schwanzfedern. *Philippia, Abh. Ber. Naturkde Mus. Kassel* 2: 182-184.

HAUKIOJA, E. 1971a. **233**
Flightlessness in some moulting passerines in Northern Europe. *Ornis Fenn.* 48: 101-116.

HAUKIOJA, E. 1971b. **234**
Summer schedule of some subarctic passerine birds with reference to postnuptial moult. *Rep. Kevo Subarctic Res. Stat.* 7: 60-69.

HAUKIOJA, E. 1971c. **235**
Processing moult card data with reference to the Chaffinch *Fringilla coelebs*. *Ornis Fenn.* 48: 25-32.

HAUKIOJA, E. and KALINAINEN, P. 1968. **236**
(On the postnuptial moult of the Willow Warbler *(Phylloscopus trochilus)*, the Whitethroat *(Sylvia communis)*, and the Meadow Pipit *(Anthus pratensis)*. *Porin Lintutiet. Yhd. Vuosik.* 2: 75-78.

HAUKIOJA, E. and KALINAINEN, P. 1972. **237**
(The ecology of some passerines during the moulting period). *Porin Lintutiet. Yhd. Vuosik.* 3: 5-16.

HAVERSCHMIDT, F. 1972. **238**
Further evidence of the 'portlandica' plumage phase of terns. *Br. Birds* 65: 117-119.

HAWTHORN, I. 1974. **239**
Moult and dispersal of juvenile Wrens. *Bird Study* 21: 88-91.

HEINROTH, O. 1906. **240**
Beobachtungen über die Schelligkeit des Federwachstums. *Orn. Monatsber.* 14: 111-115.

HEINROTH, O. 1907. **241**
Über die Schwanzmauser von *Certhia familiaris*. *J. Orn.* 55: 623-624.

HEINROTH, O. 1916. **242**
Über die Schwingenmauser junger Spechte. *J. Orn.* 64. 155-156.

HEINROTH, O. 1944. **243**
Wann stinkt der Wiedehopf? *Orn. Monatsber.* 52: 45-46.

HELMINEN, M. 1963. **244**
Composition of the Finnish populations of Capercaillie, *Tetrao urogallus*, and Black Grouse, *Lyrurus tetrix*, in the autumns of 1952-1961, as revealed by a study of wings. *Papers on Game Research*, No. 23: 1-124. Finnish Game Foundation, Helsinki.

HEREWARD, A.C. 1976. **245**
Bearded Tit moult at Radipole. *Radipole* 2 (1973-1975): 41-48.

HEREWARD, A.C. 1979. **246**
The autumn moult of the Yellow Wagtail. *Ringing and Migration* 2: 113-117.

HERRERA, C.M. 1974. **247**
El paso otoñal de *Sylvia borin* y *Sylvia communis* en la Reserva de Doñana. *Doñana, Acta Vert.* 1: 83-119.

HERROELEN, P. 1960. **248**
De rui van de Boerenzwaluw, *Hirundo rustica* L., in Belgisch-Congo. *Gerfaut* 50: 87-99.

HERROELEN, P. 1962. **249**
De Appelvink, *Coccothraustes coccothraustes* (Linné), in België. *Gerfaut* 52: 173-205.

HERROELEN, P. 1965. **250**
Het Smelleken, *Falco columbarius* Linnaeus, in België. *Gerfaut* 55: 384-393.

HERROELEN, P. 1980. **251**
Onderbroken handpennenrui bij zangvogels die over korte afstand trekken. *Ornis Brabant* 85: 13-15.

HERROELEN, P. and VAN DER MEIREN, P. 1966. **252**
Vrooge rui en onderbroken rui by de Fitis, *Phylloscopus trochilus*. *Gerfaut* 56: 416-417.

HEWSON, R. 1973. **253**
The moults of captive Scottish ptarmigan *(Lagopus mutus)*. *J. Zool., Lond.* 171: 177-187.

HILDÉN, O. and VUOLANTO, S. 1972. **254**
Breeding biology of the Red-necked Phalarope *Phalaropus lobatus* in Finland. *Ornis Fenn.* 49: 57-85.

HOFFMANN, L. 1957. **255**
Le passage d'automne du Chevalier Sylvain *(Tringa glareola)* en France méditerranéenne. *Alauda* 25: 30-42.

HŐHN, E.O. 1970. **256**
Gonadal hormone concentrations in northern phalaropes in relation to nuptial plumage. *Can. J. Zool.* 48: 400-401.

HOLMES, R.T. 1966. **257**
Molt cycle of the Red-backed Sandpiper *(Calidris alpina)* in western North America. *Auk* 83: 517-533.

HOLMES, R.T. 1971. **258**
Latitudinal differences in the breeding and molt schedules of Alaskan Red-backed Sandpipers *(Calidris alpina)*. *Condor* 73: 93-99.

HOLYOAK, D.T. 1970. **259**
Moorhen deliberately plucking its own flight-feathers. *Br. Birds* 63: 339-340.

HOLYOAK, D. 1974. **260**
Moult seasons of the British Corvidae. *Bird Study* 21: 15-20.

HOPE JONES, P. and KINNEAR, P.K. 1979. **261**

Moulting Eiders in Orkney and Shetland. *Wildfowl* 30: 109-113.

HORI, J. 1966. **262**
Moult migration of second-summer Shelduck. *Bird Study* 13: 99-100.

HULSCHER, J.B. 1977. **263**
The progress of wing-moult of Oystercatchers *Haematopus ostralegus* at Drachten, Netherlands. *Ibis* 119: 507-512.

HUME, R.A. 1976. **264**
The pattern of Mediterranean Gull records at Blackpill, West Glamorgan. *Br. Birds* 69: 503-505.

HUME, R.A. and LANSDOWN, P.G. 1974. **265**
Mediterranean Gulls at Blackpill, Glamorgan. *Br. Birds* 67: 17-24.

HUMPHREY, P.S. and CLARK, G.A. 1961. **266**
Pterylosis of the Mallard duck. *Condor* 63: 365-385.

HUMPHREY, P.S. and PARKES, K.C. 1959. **267**
An approach to the study of moult and plumages. *Auk* 76: 1-31.

HUMPHREY, P.S. and PARKES, K.C. 1963. **268**
Comments on the study of plumage succession. *Auk* 80: 496-503.

HYYTIÄ, K. and VIKBERG, P. 1973. **269**
Autumn migration and moult of the Spotted Flycatcher *Muscicapa striata* and the Pied Flycatcher *Ficedula hypoleuca* at the Signilskär bird station. *Ornis Fenn.* 50: 134-143.

INGOLFSSON, A 1970. **270**
The moult of remiges and rectrices of Great Black-backed Gulls *Larus marinus* and Glaucous Gulls *L. hyperboreus* in Iceland. *Ibis* 112: 83-92.

INSLEY, H., YOUNG, L. and DUDLEY, B. 1980. **271**
Primary moult in the Collared Dove. *Bird Study* 27: 101-107.

INSLEY, H. and JACKSON, R.V. n.d. **272**
Warbler passage at Lower Test Marsh during autumn 1975. *Hampshire Bird Report* 1975: 49-54.

JACKSON, A.C. 1915. **273**
Notes on the moults and sequence of plumages of some British ducks. *Br. Birds* 9: 34-42.

JACKSON, A.C. 1917-1919. **274**
The moults and sequence of plumage of the British waders. *Br. Birds* 11: 55-64, 83-85, 105-116, 177-182, 228-230, 12: 4-13, 39-41, 104-113, 146-151, 172-179.

JACKSON, A.C. 1918. **275**
Notes on the relation between moult and migration as observed in some waders. *Br. Birds* 11: 197-203.

JAKOBER, H. and STAUBER, W. 1980. **276**
Flügellängen und Gewichte einer südwestdeutschen Population des Neuntöters *(Lanius collurio)* unter Berücksichtigung der geschlechtsspezfischen Arbeitsteilung während der Brutperiode, *Vogelwarte* 30: 198-208.

JANY, E. 1954. **277**
Die Frühjahrsmauser der Wüstengrasmücke *(Sylvia d. deserticola). Vogelwarte* 17: 196-199.

JEPSEN, P.U. 1973. **278**
Studies of the moult migration and wing-feather moult of the Goldeneye *(Bucephala clangula)* in Denmark. *Danish Review of Game Biology* 8 (6): 1-23.

JOENSEN, A.H. 1973. **279**
Moult migration and wing-feather moult of seaducks in Denmark. *Danish Review of Game Biology* 8 (4): 1-42.

JOHNS, J.E. 1964. **280**
Testosterone-induced nuptial feathers in phalaropes. *Condor* 66: 449-455.

JOHNSON, C. 1982. **281**
The taxonomy of *Larus argentatus* and *Larus fuscus* in north-western Europe. Unpubl. Ph.D. thesis, University of Nottingham.

JOHNSON, C. and MINTON, C.D.T. 1980. **282**
The primary moult of the Dunlin *Calidris alpina* at the Wash. *Ornis Scand.* 11: 190-195.

JOHNSON, O.W. 1973. **283**
Reproductive condition and other features of shorebirds resident at Eniwetok Atoll during the boreal summer. *Condor* 75: 336-343.

JOHNSON, O.W. 1977. **284**
Plumage and moult in shorebirds summering at Enewetak Atoll. *Auk* 94: 222-230.

JOHNSTON, D.W. 1961. **285**
Timing of annual moult in the Glaucous Gulls of northern Alaska. *Condor* 63: 474-478.

JOLLIE M. 1947. **286**
Plumage changes in the Golden Eagle. *Auk* 64: 549-576.

JOUANIN, C. and ROUX, F. 1966. **287**
Scientific Expedition to the Salvage Islands. July 1963. VII. La colonie de Puffins Cendrés *Calonectris diomedea borealis* (Cory) de Selvagem Grande. *Bolm. Mus. munic. Funchal* 20: 14-28.

JOYNER, D.E. 1978. **288**
Wing molt in female Ruddy Ducks. *Condor* 80: 102-103.

JUCKWER, E.A. 1970. **289**
Die Grossgefiedermauser der Tauben am Beispiel der Türkentaube, *Streptoplia decaocto. Bonn. zool. Beitre.* 21: 237-250.

KAINADY, P.V.G. 1976. **290**
Hirundo rustica moulting remiges while rearing nestlings in Iraq. *Bulletin Basrah nat. Hist. Mus.* 3: 95-100.

KAINADY, P.V.G. 1977. **291**
The overlap of breeding and moulting in some Iraqi birds. *Bulletin Basrah nat. Hist. Mus.* 4: 65-68.

KALCHREUTER, H. 1969. **292**
Die Mauser der Rabenkrähe *(Corvus corone). Anz. orn. Ges. Bayern* 8: 578-592.

KARHU, S. 1973. **293**
On the development stages of chicks and adult Moorhens *Gallinula chloropus* at the end of a breeding season. *Ornis Fenn.* 50: 1-17.

KASPAREK, M. 1976. **294**
Über Populationsunterschiede im Mauserverhalten der Rauchschwalbe *(Hirundo rustica). Vogelwelt* 97: 121-132.

KASPAREK, M. 1979a. **295**
Suspended wing moult in a Tree Sparrow. *Ringing and Migration* 2: 158-159.

KASPAREK, M. 1979b. **296**
(The sequence of the complete postnuptial moult in the Reed Bunting *Emberiza schoeniclus*). *J. Orn.* 120: 247-264.

KASPAREK, M. 1980. **297**
(Seasonal aspects of the moult of the Reed Bunting *Emberiza schoeniclus* (L.)). *Ökol. Vogël (Ecol. Birds)* 2: 1-35.

KASPAREK, M. 1981. **298**
Die Mauser der Singvögel Europas —ein Feldführer Dachverband Deutscher Avifaunisten.

KEAR, J. 1972. **299**
Reproduction and family life. pp. 79-124 *in* SCOTT, P. and The Wildfowl Trust. *The Swans.* Michael Joseph, London.

KEAST, A. 1968. **300**
Moult in birds of the Australian dry country relative to rainfall and breeding. *J. Zool., Lond.* 155: 185-200.

KEMPER, T. 1959. **301**
Notes on the breeding cycle of the Red Crossbill *(Loxia curvirostra)* in Montana. *Auk* 76: 181-189.

KENDEIGH, S.C., DOL'NIK V.R. and GAVRILOV, V.M. 1977. Avian energetics. pp. 127-204 *in* PINOWSKI, J. and KENDEIGH, S.C. (Eds.). *Graminivorous birds in ecosystems.* Cambridge University Press, Cambridge. **302**

KENNEDY, R.J. n.d. **303**
Data from Redpolls in 1972. *Bird Ringing in SW Lancashire* 2 (1972): 13-17.

KHROKOV, V.V. 1978a. **304**
[On the moulting of some species of waders in Kazakhstan]. *Trudy Inst. Zool. Alma Ata (Ser. Zool.)* 38: 65-76.

KHROKOV, V.V. 1978b. **305**
[The biology of the Black-tailed Godwit in the Tengiz-Kaurgaldjin Depression (central Kazakhstan)]. *Trudy Inst. Zool. Alma Ata (Ser. Zool.)* 38: 29-41

KING, J.R. 1972. **306**
Postnuptial and postjuvenal molt in Rufous-collared Sparrows in northwestern Argentina. *Condor* 74: 5-16.

KING, J.R. 1974. **307**
Seasonal allocation of time and energy resources in birds. pp. 4-85 *in* PAYNTER, R.A. (Ed.). *Avian energetics.* Publ. Nuttall Ornithol. Club No. 15.

KING, J.R. 1980. **308**
Energetics of avian moult. *Acta XVII Congressus Internationalis Ornithologici, Berlin 1978:* 312-317.

KIPP, F.A. 1956. **309**
Progressive Merkmale des Jugendkleides bei den Spechten. *J. Orn.* 97: 403-410.

KIRPICHEV, S.P. 1972. **310**
[Moult of the Capercaillie *Tetrao urogallus.*]. *Ornithologiya* 10: 303-319.

KISTCHINSKI, A.A. 1975. **311**
Breeding biology and behaviour of the Grey Phalarope *Phalaropus fulicarius* in East Siberia. *Ibis* 117: 285-301.

KITTLE, T. 1975. **312**
Weights and moult of Green Sandpipers in Britain. *Ringing and Migration* 1: 52-55.

KOP, P.P.A.M. 1971. **313**
Some notes on the moult and age determination in the Great Crested Grebe. *Ardea* 59: 56-60.

KOSHELEV, A.I. 1976. **314**
(Place, time and character of moulting of the Coot *(Fulica atra)* in the Barabinsky forest-steppe (West Siberia).). *Zool. Zh.* 55: 879-884.

KOTOV, A.A. and NOSKOV, G.A. 1978. **315**
(Comparative characteristics of the moult in the Blue Hill Pigeon *Columba rupestris,* Rock Dove, *C. livia,* and domestic dove). *Zool. Zh.* 57: 1202- 1209.

KOZLOVA, E.V. 1957. **316**
[Relationship between migration and moulting periods in Palaearctic Charadriiformes]. *Trans. 2nd. Baltic Orn. Conf.:* 153-158.

KOZLOVA, E.V. 1961. **317**
Fauna of U.S.S.R. Birds Vol. II, No. 3 Charadriiformes. Suborder Alcae. Israel Program for Scientific Translations, Jerusalem.

KRAMER, G. 1950. **318**
Über die Mauser, insbesondere die sommerliche Kleingefieder-mauser beim Neuntöter *(Lanius collurio* L.). *Ornithologische Berichte* 3: 15-23.

KRÄTZIG, H. 1936. **319**
Der Frühsommerzug des Stars auf der Windenburger Ecke. *Vogelzug* 7: 1-16.

KUKISH, A.A. 1974. **320**
(The moult of White *(Motacilla alba* L.*)* and Yellow *(Motacilla flava* L.*)* Wagtails in Leningrade region.) *Vestnik Leningr. Gos. Univ. (Biol.)* 1974 (3): 20-25.

KUMARI, E. 1980. **321**
Moult and moult migration of waterfowl in the Baltic basin. *Acta Orn.* 17: 37-44.

LACHNER, R. 1965. **322**
Untersuchungen an der Türkentaube, *Streptopelia decaocto,* über Alterskennzeichen, Geschlechtsdimorphismus und Mauser. *Vogelwelt* 86: 79-95.

LANGSLOW, D.R. n.d. **323**
The timing of breeding, moult and migration of Lesser Redpolls in the Lothians. *Edinburgh Ringing Group Report* 5 (1977): 12-23.

LANGSLOW, D.R. and DA PRATO, S. n.d. **324**
Retrap rates, moult and migration of Song Thrushes and Blackbirds. *Edinburgh Ringing Group Report* 3 (1975): 26-30.

LANGSLOW, H.K. n.d. **325**
The Longniddry Linnet roost. *Edinburgh Ringing Group Report* 1 (1973): 11-20.

LEHIKOINEN, E. and LAAKSONEN, M. 1977. **326**
Frequency of tail-moult in immature Finnish Greenfinches. *Ornis Fenn.* 54: 133-134.

LEHIKOINEN, E. and NIEMELÄ, P. 1977. **327**
(Moult study on passerines). *Lintumies* 12: 33-44.

LILJA, I. 1969. **328**
(On the postnuptial wing moult of migratory Dunlin *(Calidris alpina)*.). *Porin Lintutieteelinen Yhdistys Vuosikirja 1968: 40-43.*

LOFTS, B., MURTON, R.K. and WESTWOOD, N.J. 1967. **329**
Experimental demonstration of a post-nuptial refactory period in the Turtle Dove *Streptopelia turtur. Ibis* 109: 352-358.

LUCAS, A.M. and STETTENHEIM, P.R. 1972. **330**
Avian anatomy. Integument. Agriculture Handbook 362, U.S. Dept. Agr., Washington, D.C.

LUDLOW, A.R. 1966. **331**
Body-weight changes and moult of some Palaearctic migrants in Southern Nigeria. *Ibis* 108: 129-132.

LYULEEVA, D.S. 1973. **332**
Features of Swallow biology during migration. pp. 219-272 *in* BYKHOVSKII, B.E. (Ed.). *Bird Migrations. Ecological and physiological factors.* Wiley, New York.

McNEIL, R. 1970. **333**
Hivernage et estivage d'oiseaux aquatiques Nord-Américains dans le nord-est du Vénézuela (mue, accumulation de graisse, capacité de vol et routes de migration). *L'Oiseau R.F.O.* 40: 185-302.

McNEIL, R. and CADIEUX, F. 1972. **334**
Fat content and flight range capabilities of some adult spring and fall migrant North American shorebirds in relation to migration routes on the Atlantic coast. *Naturaliste Canadien* 99: 589-606.

MACKIE, P. 1976. **335**
A short note on Redshank in the Upper Clyde estuary. *Wader Study Group Bull.* 17: 5-10.

MAINWOOD, A.R. 1974. **336**
The Dingwall Pied Wagtail roost. *Highland Ringing Group Report* 1: 13-19.

MARPLES, B.J. 1942. **337**
A study of the Little Owl, *Athene noctua,* in New Zealand. *Trans. Roy. Soc. New Zealand* 72: 237-252.

MARQUISS, M. 1980. **338**
Using moult and iris colour to age Sparrowhawks. *Ringers' Bulletin* 5: 94-95.

MARSHALL, A.J. and COOMBS, C.J.F. 1957. **339**
The interaction of environmental, internal and behavioural factors in the rook *Corvus f. frugilegus* Linneaeus. *Proc. zool. Soc. Lond.* 128: 545-598.

MARSHALL, A.J. and SERVENTY, D.L. 1956. **340**
Moult adaptation in relation to long-distance migration in petrels. *Nature* 177: 943.

MARTÍNEZ, I. 1982. **341**
Notes sobre l'anellament i muda dels Blauets *(Alcedo atthis* L.). *Butlletí del Grup Català d'Anellament* 2 (1): 5-15.

MASSEY, B.W. and ATWOOD, J.L. 1978. **342**
Plumages of the Least Tern. *Bird-Banding* 49: 360-371.

MATHIASSON, S. 1965. **343**
(Some information on weight and moult in the Wood Pigeon, *Columba palumbus*). *Göteborgs Naturhist. Mus. Yearbook* 1965: 16-23.

MATHIASSON, S. 1971. **344**
(Investigation concerning Lesser Whitethroats *(Sylvia curruca)* in the Nile valley in the Sudan). *Vogelwarte* 26: 212-221.

MATHIASSON, S. 1973. **345**
A moulting population of non-breeding Mute Swans with special reference to flight-feather moult, feeding ecology and habitat selection. *Wildfowl* 24: 43-53.

MAUSERBERGER, G. 1957. **346**
Über das Pränuptialkleid des Kampflaüfers *Philomachus pugnax* (L.). *J. Orn.* 98: 356-357.

MAWSON, G.P. 1977. **347**
Moult in the juvenile Robin. *Breck Ringing Group Report* 2: 29-31.

MAY, J.B. 1930. **348**
Simultaneous loss of primaries in prenuptial molt of Loon. *Auk* 37: 412-414.

MAYAUD, N. 1931a. 349
Contribution a l'etude de la mue des puffins. *Alauda* 3: 230-249.

MAYAUD, N. 1931b. 350
A propos de la mue prénuptiale des plongeons. *Alauda* 3: 577-578.

MAYAUD, N. 1939. 351
Sur la distinction des plumages et de l'âge chez le Gros-bec casse-noyaux *Coccothraustes coccothraustes coccothraustes* (L.). *Arch. suisses Orn.* 1: 535-539.

MAYAUD, N. 1948. 352
La mue et les plumages du Geai *Garrulus glandarius*. *Alauda* 16: 168-179.

MAYAUD, N. 1949-1950. 353
Nouvelles précisions sur la mue des Procellariens. *Alauda* 17: 144-155, 18: 222-233.

MAYAUD, N. 1952. 354
Le phylum marin d'*Anthus spinoletta* ses particularités écologiques et morphologiques. *Alauda* 20: 65-79.

MAYR, E. and MAYR, M. 1954. 355
The tail moult of small owls. *Auk* 71: 172-178.

MEAD, C. 1967. 356
Long-tailed Tit moult. *Ringers' Bulletin* 3: 11-12.

MEAD, C.J. 1975. 357
Juvenile hirundines starting primary moult in Europe. *Ringing and Migration* 1: 57.

MEAD, C.J. 1980. 358
Sand Martins moulting primaries in Britain. *Bird Study* 27: 51-53.

MEAD, C.J. and WATMOUGH, B.R. 1976. 359
Suspended moult of Trans-Saharan migrants in Iberia. *Bird Study* 23: 187-196.

MEBS, T. 1960. 360
Untersuchungen über den Rhythmus der Schwingen- und Schwanzmauser bei grossen Falken. *J. Orn.* 101: 175-194.

MEDWAY, Lord. 1973. 361
A ringing study of migratory Barn Swallows in West Malaysia. *Ibis* 115: 60-86.

MEIER, A.H. and FERRELL, B.R. 1978. 362
Avian endocrinology. pp. 213-271 *in* BUSH, A.H. (Ed.). *Chemical zoology*. Academic Press, London.

MEINERTZHAGEN, R.M. 1956. 363
Birds in Greenland. *Bull. Br. Orn. Club* 76: 17-22.

MEISE, W. 1952. 364
Über Zug und Mauser des Kiebitzregenpfeifers, *Squatarola squatarola* (L.). *Beitr. Vogelkunde* 2: 137-151.

MELDE, M. 1973. 365
Der Haubentaucher. Neue Brehm-Bücheri, Wittenberg Lutherstadt.

MELVILLE, D.S. 1981. 366
Spring measurements, weights and plumage status of *Calidris ruficollis* and *C. ferruginea* in Hong Kong. *Wader Study Group Bull.* 31: 18-21.

MENDELSOHN, J.M. 1973. 367
Some observations on age ratio, weight and moult of the European Swallow, *Hirundo rustica* L., in the Central Transvaal (Aves: Hirundinidae). *Ann. Transvaal Mus.* 28: 79-89.

MICHENER, H. and MICHENER, J.R. 1938. 368
Bars in flight feathers. *Condor* 40: 149-160.

MIDDLEMISS, E. 1961. 369
Biological aspects of *Calidris minuta* while wintering in South-west Cape. *Ostrich* 32: 107-121.

MIDDLETON, A.L.A. 1969. 370
The moult of the European Goldfinch near Melbourne, Victoria. *Emu* 69: 145-154.

MILLER, A.H. 1941. 371
The significance of moult centres among the secondary remiges in the Falconiformes. *Condor* 43: 113-115.

MILLER, A.H. 1961. 372
Molt cycles in equatorial Andean Sparrows. *Condor* 63: 143-161.

MILLER, W DeW. 1924. 373
Variations in the structure of the aftershaft and their taxonomic value. *Amer. Mus. Novit.* 140: 1-6.

MILSTEIN, P. le S., PRESTT, I. and BELL, A.A. 1970. 374
The breeding cycle of the Grey Heron. *Ardea* 58: 171-257.

MINTON, C.D.T. 1971. 375
Mute Swan flocks. *Wildfowl* 22: 71-88.

MINTON, C.D.T. 1975. 376
The waders of the Wash — Ringing and biometric studies. Wash Feasibility Study. 21pp.

MINTON, C.D.T. 1976. 377
The moult of waders. pp. 146-156 *in* GOODERS, J. (Ed.). *The third bird watchers' book*. David and Charles, Newton Abbot.

MINTON, C.D.T. and STANLEY, P.I. 1972. 378
Biometric variations in the Curlew Sandpiper. *Wader Study Group Bull.* 6: 7-9.

MONAGHAN, P. 1977. 379
The utilization of urban resources by the Herring Gull *(Larus argentatus)*. Unpubl. Ph.D. thesis, University of Durham.

MONAGHAN, P. and DUNCAN, N. 1979. 380
Plumage variation of known age Herring Gulls. *Br. Birds* 72: 100-103.

MOREAU, R.E. 1937. 381
The comparative breeding biology of the African Hornbills (Bucerotidae). *Proc. Zool. Soc. Lond.* 107A: 331-346.

MOREAU, R.E. 1972. 382
The Palaearctic-African bird migration systems. Academic Press, London.

MOREL, G. and ROUX, F. 1966. 383
Les migrateurs palearctiques au Sénégal. *Terre et Vie* 113: 19-72, 143-176.

MORLION, M.L. and VANPARIJS, P. 1979. 384
The pterylosis of five European Corvids. *Gerfaut* 69: 357-378.

MORRISON, R.I.G. 1976. 385
Moult of the Purple Sandpiper *Calidris maritima* in Iceland. *Ibis* 118: 237-246.

MOSER, M.E. (Ed.). 1981. 386
Shorebird studies in North West Morocco. Durham University Sidi Moussa Expedition, Durham

MOUNTFORT, G. 1957. 387
The Hawfinch. Collins, London.

MOYSON, I.Y. 1973. 388
Pterylography of the wing and tail of the Starling, *Sturnus vulgaris vulgaris*. *Gerfaut* 63: 257-278.

MUDGE, G.P. 1978a. 389
Observations on Glaucous Gulls. pp. 149-151 *in* GREEN, G.H. and GREENWOOD, J.J.D. (Eds.). *Joint biological expedition to North East Greenland 1974*. Dundee University N.E. Greenland Expedition, Dundee.

MUDGE, G.P. 1978b. 390
Ecological studies of Herring Gulls *(Larus argentatus* Pont.) and other Larini, in an urban environment. Unpubl. Ph.D. thesis, University College, Cardiff.

MURTON, R.K. 1968. 391
Breeding, migration and survival of Turtle Doves. *Br. Birds* 61: 193-212.

MURTON, R.K., THEARLE, R.J.P. and COOMBS, C.F.B. 1974a. 392
Ecological studies of the feral pigeon *Columba livia* var. III. Reproduction and plumage polymorphism. *J. appl. Ecol.* 11: 841-854.

MURTON, R.K., WESTWOOD, N.J. and ISAACSON, A.J. 1974b. 393
Factors affecting egg-weight, bodyweight and moult of the Woodpigeon *Columba palumbus*. *Ibis* 116: 52-73.

MURTON, R.K. and WESTWOOD, N.J. 1977. 394
Avian breeding cycles. Clarendon Press, Oxford.

MUZAEV, V. 1978. 395
(The moult of the Garden Warbler *Sylvia borin* in the northeast Leningrad Region). *Vest. Leningradsk. Univ., Biol. Div.* 1978: 7-11.

MYRCHA, A. and PINOWSKI, J. 1970. 396
Weights, body composition, and caloric value of postjuvenal molting European Tree Sparrows *(Passer montanus)*. *Condor* 72: 175-181.

NADLER, T. 1978. **397**
Zur Mauser der Zwergseeschwalbe *(Sterna albifrons). Zoologische Abhandlungen, Staatliches Museum für Tierkunde in Dresden* 35 (15): 253-260.

NAU, B.S. 1967. **398**
Further studies of the Greenfinch *(Chloris chloris)* at Rye Meads. *Report of the Rye Meads Ringing Group* 4: 26-34.

NELSON, J.B. 1978a. **399**
The Sulidae. Oxford University Press, Oxford.

NELSON, B. 1978b. **400**
The Gannet. Poyser, Berkhamstead.

NETHERSOLE-THOMPSON, D. 1966. **401**
The Snow Bunting. Oliver and Boyd, Edinburgh.

NEWTON, A. 1900. **402**
The Great Shearwaters in Scottish waters. *The Annals of Scottish Natural History* 33: 142-147.

NEWTON, I. 1966. **403**
The moult of the Bullfinch *Pyrrhula pyrrhula. Ibis* 108: 41-67.

NEWTON, I. 1967. **404**
Feather growth and moult in some captive finches. *Bird Study* 14: 10-24.

NEWTON, I. 1968a. **405**
The moulting seasons of some finches and buntings. *Birds Study* 15: 84-92.

NEWTON, I. 1968b. **406**
The temperatures, weights, and body composition of molting Bullfinches. *Condor* 70: 323-332.

NEWTON, I. 1969. **407**
Moults and weights of captive Redpolls *Carduelis flammea. J. Orn.* 110: 53-61.

NEWTON, I. 1970. **408**
Irruptions of Crossbills in Europe. pp. 337-357 *in* WATSON, A. (Ed.). *Animal populations in relation to their food resources.* Blackwell, Oxford.

NEWTON, I. 1972. **409**
Finches. Collins, London.

NEWTON, I. and MARQUISS, M. 1982. **410**
Moult in the Sparrowhawk. *Ardea* 70: 163-172.

NIEBOER, E. 1972. **411**
Preliminary notes on the primary moult in Dunlins *Calidris alpina. Ardea* 60: 112-119.

NIETHAMMER, G. 1970a. **412**
Zur Mauser der Ringeltaube *(Columba palumbus). J. Orn.* 111: 367-377.*

NIETHAMMER, G. 1970b. **413**
Mauserzug des Kiebitzes? *Vogelwarte* 25: 331-334.

NITZSCH, C.L. 1867. **414**
Pterylography. (SCLATER, P.L., Ed.). Ray Society, London.

NORMAN, S.C. 1979. **415**
A study of the post juvenile moult and adult moult of the Willow Warbler. *South Cleveland Ringing Group Report* 1: 13-28.

NORMAN, S.C. 1981. **416**
A study of post-juvenile moult in Willow Warblers. *Ringing and Migration* 3: 165-172.

NOSKOV, G.A. 1975. **417**
(Moulting in Chaffinch, *Fringilla coelebs*). *Zool. Zh.* 54: 413-424.

NOSKOV, G.A. and GAGINSKAYA, A.R. 1972. **418**
(Methods of describing the state of moulting in birds). *Communications of the Baltic Commission for the Study of Bird Migration* 7: 154-163.

OAG MÜNSTER. 1975. **419**
(Migration, moult and biometry of the Snipe *(Gallinago gallinago)* in the sewage farms of Münster (52°04'N 7°41'E).). *J. Orn.* 116: 455-487.

OELKE, H. 1969. **420**
Die Brandgans *(Tadorna tadorna)* im Mausergebiet Grosser Knechtsand. *J. Orn.* 110: 170-175.

OGILVIE-GRANT, W.R. 1910. **421**
Exhibition of a series of examples of the Red Grouse *(Lagopus scoticus),* with remarks on their seasonal changes of plumage. *Bull. Br. Orn. Club* 25: 122-124.

OLIVER, P.J. 1980. **422**
Moulting Tufted Ducks and Pochards in the London area. *London Bird Report* 44: 80-84.

OPDAM, P. and MÜSKENS, G. 1976. **423**
Use of shed feathers in population studies of *Accipiter* hawks (Aves, Accipitriformes, Accipitridae). *Beaufortia* 24 (312): 55-62.

ORELL, M. and OJANEN, M. 1980. **424**
Overlap between breeding and moulting in the Great Tit *Parus major* and Willow Tit *P. montanus* in northern Finland. *Ornis Scand.* 11: 43-49.

ORING, L.W. 1968. **425**
Growth, molts and plumages of the Gadwall. *Auk* 85: 355-380.

OWEN, M. and KING, R. 1979. **426**
The duration of the flightless period in free-living Mallard. *Bird Study* 26: 267-269.

OWEN, M. and OGILVIE, M.A. 1979. **427**
Wing molt and weights of Barnacle Geese in Spitsbergen. *Condor* 81: 42-52.

PAGE, G. 1974. **428**
Age, sex, molt and migration of Dunlins at Bolinas Lagoon. *Western Birds* 5: 1-12.

PALMER, R.S. 1941. **429**
'White-faced' terns. *Auk* 58: 164-178.

PALMER, R.S. (Ed.) 1962, 1976a, 1976b. **430**
Handbook of North American Birds. Vols. 1,2,3. Yale University Press, New Haven.

PALMER, R.S. 1972. **431**
Patterns of molting. pp. 65-102 *in* FARNER, D.S. and KING, J.R. (Eds.). *Avian Biology* II. Academic Press, London.

PALUDAN, K. 1936. **432**
Report on the birds collected during Professor O. Olufsen's expedition to French Sudan and Nigeria in the year 1927; with field notes by the collector Mr. Harry Madsen. *Vidensk. Medd. fra Dansk naturh. Foren.* 100: 247-346.

PARKES, K.C. 1966. **433**
Speculations on the origin of feathers. *Living Bird* 5: 77-86.

PARMELEE, D.F. 1968. **434**
Plectrophenax nivalis nivalis (Linnaeus) Snow Bunting. pp. 1652-1675 *in* AUSTIN, O.L. (Ed.). Life histories of North American Cardinals, Grosbeaks, Buntings, Towhees, Finches, Sparrows, and Allies. Order Passeriformes: Family Fringillidae. *U.S. National Museum Bulletin* 237.

PARR, R. 1975. **435**
Aging Red Grouse chicks by primary molt and development. *J. Wildl. Mgmt.* 39: 188-190.

PATON, D.C., WYKES B.J. and **436**
DANN, P. 1982. Moult of juvenile Curlew Sandpipers in southern Australia. *Emu* 82: 54-56.

PAULUSSEN, J. 1980. **437**
(Spring head-moult of the Black-headed Gull). *De Wielewaal* 46: 297-300.

PAYNE, R. 1969. **438**
Overlap of breeding and molting schedules in a collection of African birds. *Condor* 71: 140-145.

PAYNE, R.B. 1972. **439**
Mechanisms and control of molt. pp. 103-155 *in* FARNER, D.S. and KING, J.R. (Eds.). *Avian Biology.* Vol. II. Academic Press, London.

PEARSON, D.J. 1972. **440**
The wintering and migration of Palaearctic passerines at Kampala southern Uganda. *Ibis* 114: 43-60.

PEARSON, D.J. 1973a. **441**
Moult of some Palaearctic warblers wintering in Uganda. *Bird Study* 20: 24-36.

PEARSON, D.J. 1973b. **442**
On the wintering of the Sprosser *Luscinia luscinia* and the Marsh Warbler *Acrocephalus palustris* in Kenya. *East Africa nat. Hist. Soc. Bull.* 1973: 140.

PEARSON, D. 1974. **443**

The timing of wing moult in some Palaearctic waders wintering in East Africa. *Wader Study Group Bull.* 12: 6-12.

PEARSON, D.J. 1975. **444**
Moult and its relation to eruptive activity in the Bearded Reedling. *Bird Study* 22: 205-227.

PEARSON, D.J. 1977. **445**
The first year moult of the Common Sandpiper *Tringa hypoleucos* in Kenya. *Scopus* 1: 89-94.

PEARSON, D.J. 1981. **446**
The wintering and moult of Ruffs *Philomachus pugnax* in the Kenyan rift valley. *Ibis* 123: 158-182.

PEARSON, D.J. and BACKHURST, G.C. 1976. **447**
The southward migration of Palaearctic birds over Ngulia Kenya. *Ibis* 118: 78-105.

PEARSON, D.J., BACKHURST, G.C. and **448**
BACKHURST, D.E.G. 1979. Spring weights and passage of Sedge Warblers *Acrocephalus schoenobaenus* in central Kenya. *Ibis* 121: 8-19.

PENNYCUICK, C.J. 1972. **449**
Animal flight. Studies in Biology No. 33. Arnold, London.

PENNYCUICK, C.J. 1975. **450**
Mechanics of flight. pp. 1-75 *in* FARNER, D.S. and KING, J.R. (Eds.). *Avian Biology*. Vol. V. Academic Press, London.

PEPLER, G.R.M. n.d. **451**
Moult records at Radipole,1972. *Radipole* 1 (1970-72): 32.

PEPLER, P. and PEPLER, G. n.d. **452**
Sedge Warblers and Reed Warblers at Radipole, 1972. *Radipole* 1 (1970-72): 11-29.

PERRINS, C. 1979. **453**
British Tits. Collins, London.

PERSSON, C. 1977. **454**
The early stages of postnuptial moult in the White Wagtail *Motacilla alba*. *Ornis Scand.* 8: 97-99.

PETTINGILL, O.S. 1970. **455**
Ornithology in laboratory and field. 4th. Ed. Burgess, Minneapolis.

PIECHOCKI, R. 1956. **456**
Über die Mauser eines gekäfigten Turmfalken *(Falco tinnunculus). J. Orn.* 97: 301-309.

PIECHOCKI, R. 1961. **457**
Über die Grossgefieder-Mauser von Schleiereule und Waldkauz. *J. Orn.* 102: 220-225.*

PIECHOCKI, R. 1963a. **458**
Über die Mauser eines gekäfigten Baumfalken *(Falco subbuteo). Beitre. Vogelkunde.* 9: 69-77.

PIECHOCKI, R. 1963b. **459**
Vorläufiges über die Mauser der Handschwingen beim Mäusebussard *(Buteo buteo). J. Orn.* 104: 182-184.

PIECHOCKI, R. 1968a. **460**
Über die Grossgefieder-Mauser einer gekäfigten Waldohreule *(Asio otus). Beitr. Vogelkunde.* 13: 455-460.

PIECHOCKI, R. 1968b. **461**
Die Grossgefieder-Mauser des Steinkauzes *(Athene noctua). J. Orn.* 109: 30-36.*

PIECHOCKI, R. 1974. **462**
(The moult of wing and tail of a caged pair of Barn Owls). *J. Orn.* 115: 436-444.

PIENKOWSKI, M.W. 1975. **463**
Studies on waders and terns caught in Morocco. pp. 62-87 *in* PIENKOWSKI, M.W. (Ed.). *Studies on coastal birds and wetlands in Morocco 1972*. The joint report of the University of East Anglia Expedition to Tarfaya Province, Morocco 1972 and the Cambridge Sidi Moussa Expedition 1972. Norwich.

PIENKOWSKI, M.W. and DICK, W.J.A. 1975. **464**
The migration and wintering of Dunlin *Calidris alpina* in north-west Africa. *Ornis Scand.* 6: 151-167.

PIENKOWSKI, M.W. and EVANS, P.R. 1979. **465**
The origins of Shelducks moulting on the Forth. *Bird Study* 26: 195-196.

PIENKOWSKI, M,W. and KNIGHT, P.J. 1975. **466**
Wing-moult of Oystercatchers in Iceland. *Ibis* 117:114.

PIENKOWSKI, M.W., KNIGHT, P.J., **467**
STANYARD, D.J. and ARGYLE, F.B. 1976. The primary moult of waders on the Atlantic coast of Morocco. *Ibis* 118: 347-365.

PIMM, S.L. 1970. **468**
Swallows in wing-moult in Southern Spain. *Bird Study* 17: 49-51.

PIMM, S.L. 1972. **469**
Moult in the Swallow. *Bird Study* 19: 116.

PIMM, S.L. 1973. **470**
The molt of the European Whitethroat. *Condor* 75: 386-391.

PIMM, S.L. 1976. **471**
Estimation of the duration of bird molt. *Condor* 78: 550.

PINOWSKI, J. 1968. **472**
Fecundity, mortality, numbers and biomass dynamics of a population of the Tree Sparrow *(Passer montanus* L.*) Ekol. Polska* A, 16: 1-58.

PIPER, S.E. 1974. **473**
Moult of the European Swallow. *Safring News* 3: 24-32.

PITELKA, F.A. 1945. **474**
Pterylography, molt, and age determination of American jays of the genus *Aphelocoma*. *Condor* 47: 229-260.

PITELKA, F.A. 1958. **475**
The timing of molt in Steller Jays of the Queen Charlotte Islands, British Columbia. *Condor* 60: 38-49.

PONCY, R. 1908. **476**
A propos de la mue des rémiges de la Mouette Rieuse *(Larus ridibundus). Bull. Soc. zool. de Genève* 1: 144-147.

POOR, H.H. 1946. **477.**
Plumage and soft-part variations in the Herring Gull. *Auk* 63: 135-151.

POSLAWSKI, A.N. 1969. **478**
Zug und Mauser des Grossen Brachvogels. *Falke* 16: 184-188.

POTTS, G.R. 1971. **479**
Moult in the Shag *Phalacrocorax aristotelis*, and the ontogeny of the "staffelmauser". *Ibis* 113: 298-305.

POULSEN, H. 1950. **480**
Observations on colour-change in the male Reed-Bunting *(Emberiza s. schoeniclus* (L.)*). Dansk orn. Foren. Tidssk.* 44: 96-99.

PRATER, T. and WILSON, J. 1972. **481**
Aspects of spring migration of Knot in Morecambe Bay. *Wader Study Group Bull.* 5: 9-11.

PRATER, A.J., MARCHANT, J.H. and **482**
VUORINEN, J. 1977. *Guide to the identification and ageing of Holarctic Waders. B.T.O. Guide 17.* British Trust for Ornithology, Tring.

PRICE, T. 1981. **483**
The ecology of the Greenish Warbler *Phylloscopus trochiloides* in its winter quarters. *Ibis* 123: 131-144.

RAKHILIN, V.K. 1973. **484**
[Notes on the moult of waders in the Far East]. pp. 132-134 *in* FLINT, V.E. (Ed.). [*Fauna and Ecology of Waders*]. Moscow Society of Naturalists, Moscow.

RAMSAY, L.N.G. 1913. **485**
On the moulting of the Herring Gull *(Larus argentatus)* and other species. *Scot. Nat.* 1913: 100-105.

RAWLES, M.E. 1960. **486**
The integumentary system. pp. 189-240 *in* MARSHALL, A.J. (Ed.) *Biology and comparative physiology of birds*. Vol. 1. Academic Press, London.

REYNOLDS, C.M. 1976. **487**
Measurements of waders less frequently ringed in Britain. 1. Little Ringed Plover *Charadrius dubius*. *Wader Study Group Bull.* 19: 10-12.

RICHARDS, B.A. and GOODWIN, D. 1950. **488**
Swallow moulting remiges in Britain. *Br. Birds* 43: 300-301.

RICHARDS, P.R. 1976. **489**
The onset of moult in the Rook. *Bird Study* 23: 212.

RICHTER, A. 1972. **490**
Zum Umfang der Jugendmauser am Flügel der Amsel *Turdus merula*. *Orn. Beob.* 69: 1-16.*

RICHTER, H. 1954. **491**
Zur Mauser der Wasseramsel, *Cinclus c. aquaticus* (Bechstein). *Beitr. Vogelkunde* 3: 251-258.

RIDDIFORD, N. 1981. **492**
Cautionary notes on ageing Redwings and Stonechats. *Ringers' Bulletin* 5: 120.

RINTOUL, L.J. and BAXTER, E.V. 1914. **493**
Notes on some passerine birds found migrating in moult. *Scot. Nat.* 1914: 245-252.

ROGGE, D. 1966. **494**
Ein Beitrag zur Mauser des Rotkehlchens *(Erithacus rubecula rebecula* L.). *Beitr. Vogelkunde* 12: 162-188.

ROHNER, C. 1981. **495**
Biometrie, Alters- und Geschlechts-merkmale des Girlitz *Serinus serinus. Orn. Beob.* 78: 1-11.

ROLLS, J.C. and SIMMONS, K.E.L. 1977. **496**
Great Crested Grebes retaining nuptial plumes throughout the year. *Br. Birds* 70: 75-76.

ROSENE, W. 1969. **497**
The Bobwhite Quail. Its life and management. Rutgers University Press, New Brunswick.

ROWAN, M.K. 1962. **498**
Mass mortality among European Common Terns in South Africa in April-May 1961. *Br. Birds* 55: 103-114.

RUGE, K. 1969. **499**
Zur Biologie des Dreizehenspechts *Picoides tridactylus* L. 2. Beobachtungen zur Mauser. *Orn. Beob.* 66: 42-54.

SACH, G. 1967. **500**
Brutbeobachtungen an Rohrweihen, *Circus aeruginosus*, im Rantum-Becken (Sylt) aus dem Jahr 1966. *Corax* 2: 9-17.

SACH, G. 1968. **501**
Die Mauser des Grossen Brachvogels, *Numenius arquata. J. Orn.* 109: 485-511.*

SACH, G. 1970. **502**
Zur Handschwingen-Mauser von *Numenius arquata. J. Orn.* 111: 105-107.

SAGE, B.L. 1968§ **503**
Ornithological transects in the North Atlantic. *Ibis* 110: 1-16.

SALOMONSEN, F. 1939. **504**
Moults and sequence of plumages in the Rock Ptarmigan *(Lagopus mutus* (Montin)*). Vidensk. Medd. Dansk Naturhist. Forening* 103: 1-491.

SALOMONSEN, F. 1941. **505**
The Black-winged Guillemot *(Uria grylle* mut. *motzfeldi* Benicken*).* (Miscellaneous notes on Greenland Ornithology V.). *Meddelelser Om Grønland* 131 (6): 1-21.

SALOMONSEN, F. 1944. **506**
The Atlantic Alcidae. The seasonal and geographic variation of auks inhabiting the Atlantic Ocean and the adjacent waters. *Medd. Göteborgs K. Vetensk.-o Vitter Samh. Handl., Sjatte Földer.*, ser. B Band 3, No. 5.

SALOMONSEN, F. 1945. **507**
Notes on the variation and moult in the Willow-Warbler *(Phylloscopus trochilus). Arkiv för Zoologi* 36A, No. 17: 1-13.

SALOMONSEN, F. 1967. **508**
Migratory movements of the Arctic Tern *(Sterna paradisaea* Pontoppoidan) in the Southern Ocean. *Dan. Biol. Medd.* 24 (1): 1-42.

SALOMONSEN, F. 1968. **509**
The moult migration. *Wildfowl* 19: 5-24.

SALOMONSEN, F. 1972. **510**
Zoogeographical and ecological problems in arctic birds. *Proc. XVth. Int. Orn. Congr., The Hague, 1970:* 25-77.

SALOMONSEN, F. 1976. **511**
The South Polar Skua *Stercorarius maccormicki* Saunders in Greenland. *Dansk. orn. Foren. Tidsskr.* 70: 81-89.

SCHERRER, B. 1972. **512**
Migration et autres types de displacements de la Mesange Noire *Parus ater* en transit au Col de la Goleze. II. *Terre et Vie* 26: 257-313.*

SCHERZINGER, W. 1974. **513**
Zur Ethologie und Jugendentwicklung der Schnee-Eule *(Nyctea scandiaca)* nach Beobachtungen in Gefangenschaft. *J. Orn.* 115: 8-49.

SCHIEMENZ, H. 1958. **514**
Zur Mauser des Kleingefieders beim Hühnerhabicht *(Accipiter gentilis* L.*). J. Orn.* 99: 59-66.

SCHIÖLER, E.L. 1921. **515**
A short description of the sequence of plumages in some palaearctic surface-feeding ducks. *Br. Birds* 15: 130-138.

SCHMITT, M.B. and WHITEHOUSE, P.J. 1976. **516**
Moult and mensural data of Ruff on the Witwatersrand. *Ostrich* 47: 179-190.

SCHMUTZ, J.K. and SCHMUTZ, S.M. 1975. **517**
Primary moult in *Circus cyaneus* in relation to nest brood events. *Auk* 92: 105-110.

SCHÖNFELD, M. 1978. **518**
Der Weidenlaubsänger. Neue Brehm-Bücheri, Wittenberg Lutherstadt.

SCHÖNFELD, M. and PIECHOCKI, R. 1974. **519**
(On the wing and tail moult of the Barn Owl *(Tyto alba guttata)*.). *J. Orn.* 115: 418-435.

SCHÜSTER, L. 1926. **520**
Beobachtungen an den Horsten der Kornweihe *Circus cyaneus cyaneus* (L.) und der Wiesenweihe *Circus pygargus* (L.). *Beitr. Fortpfl. Biol. Vögel* 2: 127-136, 145-151.

SCHÜSTER, L. 1928. **521**
Einige brutbiologische Beobachtungen aus dem Jahre 1928. *Beitr. Fortpfl. Biol. Vögel* 4: 209-214.

SCHUSTER, S. and BLUM, V. 1972. **522**
Der Bodensee als Mauserstation auf dem Fruhjahrszug des Kampfläuders *(Philomachus pugnax)? Anzeiger orn. Ges. Bayern* 11: 36-42.

SCHÜZ, E. 1936. **523**
Von der Mauser des Polar-Seetauchers *(Colymbus arcticus). Orn. Mber.* 44: 65-71.

SCHÜZ, E. 1974. **524**
Über den Zug von *Gavia arctica* in der Paläarktis. *Ornis Fenn.* 51: 183-194.*

SCOTT, D.A. 1970. **525**
The breeding biology of the Storm Petrel, *Hydrobates pelagicus.* Unpubl. D.Phil. thesis, University of Oxford.

SCOTT, R.E. 1965. **526**
First-year Starling retaining juvenile flight feathers and comments on post-fledging moult. *Bull. Br. Orn. Club* 85: 66-67.

SCOTT, R.E. and GRANT, P.J. 1969. **527**
Uncompleted moult in *Sterna* terns and the problems of identification. *Br. Birds* 62: 93-97.

SEEL, D.C. 1976. **528**
Moult in five species of Corvidae in Britain. *Ibis* 118: 491-536.

SEEL, D.C. 1980. **529**
Geographical distribution and moult of the Cuckoo *Cuculus canorus* in the Western Palaearctic and Ethiopian regions. *Acta XVII Congressus Internationalis Ornithologici, Berlin, 1978:* 1399.

SERVHEEN, C. 1976. **530**
Deck-feather molt in Bald and Golden Eagles in relation to feather mounting of radio transmitters. *Raptor Res.* 10: 58-60.

SHEWELL, E.L. 1959. **531**
The waterfowl of Barberspan. *Proc. Pan-African Orn. Congr. Ostrich Suppl.* 3: 160-179.

SHORTEN, M. 1974. **532**
The European Woodcock *(Scolopax rusticola)* a search of the literature since 1940. *Report 21.* Game Conservancy, Fordingbridge.

SIBLEY, C.G. 1957. **533**
The abbreviated inner primaries of nestling woodpeckers. *Auk* 74: 102-103.

SIEGFRIED, W.R. 1965. **534**
Times of wing-moult in captive pen-reared Red-billed Teal and Egyptian Geese. *Ostrich* 36: 94.

SIEGFRIED, W.R. 1967. **535**
Trapping and ringing of Egyptian Geese and African Shelduck at Vogelvlei, Cape. *Ostrich* 38: 173-178.

SIEGFRIED, W.R. 1973. **536**
Wing moult of Ruddy Ducks in Manitoba. *Bull. Br. Orn. Club* 93: 98-99.

SIEWERT, H. 1930. **537**

Bilder aus dem Leben eines Sperberpaars zur Brutzeit. *J. Orn.* 78: 245-254.

SIMMONS, K.E.L. 1964. **538**
Feather maintenance. pp. 278-286 *in* THOMSON, A.L. (Ed.). *A new dictionary of birds.* Nelson, London.

SIMMONS, K.E.L. 1974. **539**
Adaptations in the reproductive biology of the Great Crested Grebe. *Br. Birds* 67: 413-437.

SKEAD, D.M. and SKEAD, C.J. 1970. **540**
Hirundinid mortality during adverse weather, November 1968. *Ostrich* 41: 247-251.

SMALLEY, F.W. 1915. **541**
Further notes on the moults and sequences of plumages in some British ducks. *Br. Birds* 9: 137-141.

SNOW, D.W. 1965. **542**
The moult enquiry. Fourth Report, June 1965. *Bird Study* 12: 135-142.

SNOW, D.W. 1967. **543**
A guide to moult in British birds. B.T.O. Field Guide No. 11. British Trust for Ornithology.

SNOW, D.W. 1969. **544**
The moult of British thrushes and chats. *Bird Study* 16: 115-129.

SNOW, D. and SNOW, B. 1976. **545**
Post-breeding moult of the Lapwing. *Bird Study* 23: 117-120.

SOKOLOVSKI, J. 1960. **546**
The Mute Swan in Poland. *State Council for Nature Conservation, Warsaw* 1: 1-28.

SONDELL, J. 1977. **547**
(Moult in the Reed Bunting *Emberiza schoeniclus* at Kvismaren). *Vår Fågelvärld* 36: 174-184.

SPAANS, A.L. 1979. **548**
Wader studies in Surinam, South America. *Wader Study Group Bull.* 25: 32-37.

SPAANS, A.L. 1980. **549**
Biometrics and moult of Sanderlings *Calidris alba* during the autumn in Suriname. *Wader Study Group Bull.* 27: 33-35.

SPENCER, B. and MEAD, C. 1978a. **550**
Hints on ageing and sexing throughout the year. *Ringers' Bulletin* 5: 38-42.

SPENCER, B. and MEAD, C. 1978b. **551**
Hints on ageing and sexing throughout the year (Part 2). *Ringers' Bulletin* 5: 55-56.

SPENCER, B. and MEAD, C. 1979. **552**
Hints on ageing and sexing (Part 3). The common warblers. *Ringers' Bulletin* 5: 63-72.

SPITZER, G. 1972. **553**
(Seasonal aspects of the biology of the Bearded Tit *(Panurus biarmicus)*). *J. Orn.* 113: 241-275.

SPOFFORD, W.R. 1946. **554**
Observations on two Golden Eagles. *Auk* 63: 85-87.

STAINTON, J.M. 1982. **555**
Timing of bathing, dusting and sunning. *Br. Birds* 75: 65-86.

STANLEY, P. 1972. **556**
The W.W.R.G. visit to Denmark and Sweden, August 1972. New information on the timing of Dunlin wing moult. *Wader Study Group Bull.* 7: 9-10.

STANLEY, P.I., BROUGH, T., FLETCHER, M.R., **557**
HORTON, N. and ROCHARD, J.B.A. 1981. The origins of Herring Gulls wintering inland in south-east England. *Bird Study* 28: 123-132.

STANYARD, D.J. 1974. **558**
A note on the primary moult of *Calidris minuta* Little Stint in Botswana. *Wader Study Group Bull.* 12: 13-14.

STEGMANN, B. 1956. **559**
Über die Herkunft des flüchtigen rosenroten Federpigments. *J. Orn.* 97: 204-205.

STEGMANN, B. 1962. **560**
Die verkümmerte distale Handschwinge des Vogelflügels. *J. Orn.* 103: 50-85.

STEINBACHER, J. 1955. **561**
Über die Schwanzmauser der Eulen (Strigidae) und Nachtschwalben (Caprimulgidae). *Senck. biol.* 36: 235-240.

STEINER, H.M. 1970. **562**
Die vom Schema der Passeres abweichende Handschwingenmauser des Rohrschwirls *(Locustella luscinioides). J. Orn.* 111: 230-236.*

STEINER, H.M. 1971. **563**
Entwicklung und Mauser der Sprinalflur bei der Bartmeise *(Panurus biarmicus). J. Orn.* 112: 36-42.*

STETTENHEIM, P. 1972. **564**
The integument of birds. pp. 2-63 *in* FARNER, D.S. and KING, J.R. (Eds.). *Avian biology.* Vol. II. Academic Press, London.

STEVENTON, D.J. 1977. **565**
Dunlin in Portsmouth, Langstone and Chichester harbours. *Ringing and Migration* 1: 141-147.

STONE, W. 1897. **566**
On the annual moult of the Sanderling. *Proc. Ac. Nat. Sc. of Philadelphia* 1897: 368-372.

STREICHERT, J. 1982. **567**
Biometrie und Handschwingenmauser der auf dem Grossen Knechtsand rastenden Zwergseeschwalben *(Sterna albifrons). Beitr. Naturk. Niedersachsens* 35: 41-53.

STRESEMANN, E. 1920. **568**
Avifauna Macedonica. Dultz, München.

STRESEMANN, E. 1921. **569**
Handschwingenmauser der *Alcedinidae. Orn. Monatsber.* 29: 37-40.

STRESEMANN, E. 1940. **570**
Zeitpunkt und Verlauf der Mauser bei einiger Entenarten. *J. Orn.* 88: 288-333.

STRESEMANN, E. 1944. **571**
Die Mauser der Wiedehopfs *(Upupa e. epops). Orn. Monatsber.* 52: 121-122.

STRESEMANN, E. 1955. **572**
Die Wanderungen des Waldlaubsängers *(Phylloscopus sibilatrix). J. Orn.* 96: 153-167.

STRESEMANN, E. 1963a. **573**
The nomenclature of plumages and molts. *Auk* 80: 1-8.

STRESEMANN, E. 1963b. **574**
Variations in the number of primaries. *Condor* 65: 449-459.

STRESEMANN, E. 1963c. **575**
Zeitraum und Verlauf der Hanschwingen-Mauser palaearktischer Möwen, Seeschwalben und Limicolen. *J. Orn.* 104: 424-435.

STRESEMANN, E. 1965. **576**
Die Mauser der Hühnervogel. *J. Orn.* 106: 58-64.

STRESEMANN, E. 1967. **577**
Inheritance and adaptation in moult. *Proc. XIVth Int. Orn. Congr., Oxford, 1966:* 75-80.

STRESEMANN, E. 1971. **578**
(On the start of primary-moult in Gulls and its release). *Vogelwarte* 26: 232-238.

STRESEMANN, E. and STEPHAN, B. 1968 **579**
Über das Remicle. *J. Orn.* 109: 315-322.

STRESEMANN, E. and STRESEMANN, V. 1965. **580**
Die Mauser des Kampfläufers *Philomachus pugnax. L'Oiseau et R.F.O.* 35 No. Special: 125-129.

STRESEMANN, E. and STRESEMANN, V. 1966. **581**
Die Mauser der Vögel. *J. Orn.* 107 Sonderheft.

STRESEMANN, E. and STRESEMANN, V. 1968a. **582**
(Moult and winterquarters of *Sylvia communis*). *J. Orn.* 109: 303-314.

STRESEMANN, E. and STRESEMANN, V. 1968b. **583**
Im Sommer mausernde Populationen der Rauchschwalbe, *Hirundo rustica. J. Orn.* 109: 475-484.*

STRESEMANN, E. and STRESEMANN, V. 1970a. **584**
Über die Vollmauser des Rohrschwirls *(Locustella luscinioides). J. Orn.* 111: 237-239.*

STRESEMANN, E. and **585**
STRESEMANN, V. 1970b. Über Mauser und Zug von *Puffinus gravis. J. Orn.* 111: 378-393.

STRESEMANN, E. and STRESEMANN, V. 1970c. **586**
Die Vollmauser der Schneeammer *Plectrophenax nivalis. Beitr. Vogelkunde.* 16: 386-392.

STRESEMANN, V. 1963. **587**
(Reversed wing- and tail-moult in *Muscicapa striata*). *J. Orn.* 104: 101-111.

STRESEMANN, V. and STRESEMANN, E. 1960. **588**
Die Handschwingenmauser der Tagraubvögel. *J. Orn.* 101: 373-403.

STRESEMANN, V. and STRESEMANN, E. 1961a. **589**
Die Handschwingen-Mauser der Kuckucke (Cuculidae). *J. Orn.* 102: 317-352.

STRESEMANN, V. and STRESEMANN, E. 1961b. **590**
Die Handschwingen-Mauser der Eisvögel (Alcedinidae). *J. Orn.* 102: 439-455.*

SULLIVAN, J.O. 1965. **591**
"Flightlessness" in the Dipper. *Condor* 67: 535-536.

SULTANA, J. and GAUCI, C. 1973. **592**
The Cetti's Warbler *Cettia cetti* in Malta. *Il-Merill* 12: 7-9.

SUMMERS, D.D.B. 1972. **593**
Pterylography, plumage development and moult of Japanese Quail *Coturnix c. japonica* in captivity. *Ibis* 114: 79-88.

SUMMERS, R.W. 1974. **594**
Arctic Ringed Plover in Eastern Scotland. *Wader Study Group Bull.* 13: 13-14.

SUMMERS, R.W. 1980. **595**
On the rate of change of moult scores in waders. *Wader Study Group Bull.* 28: 24.

SUMMERS, R.W. 1981. **596**
Estimates of the duration of the primaryy moult of the Redshank. *Wader Study Group Bull.* 33: 5.

SUMMERS, R., ATKINSON, N. and NICOLL, M. n.d. **597**
Aspects of Turnstone ecology in Scotland. *Tay Ringing Group Report* 1974: 3-10.

SUMMERS, R.W. SWANN, R.L. and NICOLL, M. 1980. Unbending moult data. *Wader Study Group Bull.* 30: 12-13. **598**

SUMMERS, R.W., SWANN, R.L. and NICOLL, M. in prep. The effect of methods on estimates of the duration of the primary moult of the Redshank. **599**

SUMPER, M. 1964. **600**
Der Durchzug des Tüpfelsumpfhuhns *(Porzana porzana)* im Ismaninger Teichgebiet bei München. *Anz. orn. Ges. Bayern* 7: 90-99.

SUTTER, E. 1941. **601**
Beitrag zur Kenntnis der postembryonalen Entwicklung des Wendehalses *(Jynx t. torquilla L.). Schweiz. Arch. Orn.* 1: 481-508.

SUTTER, E. 1951. **602**
Zur Mauser des Roten Milans. *Orn. Beob.* 48: 166-167.

SUTTER, E. 1956. **603**
Zur Flügel- und Schwanzmauser des Turmfalken *(Falco tinnunculus). Orn. Beob.* 53: 172-182.

SUTTER, E. 1971. **604**
Ausbildung und Mauser des Flügelgefieders beim juvenilen Jagdfasan *Phasianus colchicus. Orn. Beob.* 68: 179-222.

SUTTER, E. 1972. **605**
Duration and pattern of postjuvenile wing moult in Black Grouse and Pheasant. *Proc. Int. Orn. Congr.* 15: 692.

SUTTER, E. 1975. **606**
Zur Grossgefiedermauser der Schleiereule. *Orn. Beob.* 72: 199-202.

SUTTER, G. 1974. **607**
Längenverhältnisse der juvenilen Handschwingen bei europäischen und indomalaiischen Spechten (Picidae). *Revue suisse Zool.* 81: 684-689.

SVENSSON, L. 1975. **608**
Identification Guide to European passerines. (2nd. Rev. Ed.). Naturhistoriska Riksmuseet, Stockholm.

SWANN, R.L. 1981. **609**
Bar-tailed Godwits on the Moray Firth. *Highland Ringing Group Report* 4: 3-5.

SWANN, R.L. n.d.a. **610**
Siskins and Redpolls at Drumnadrochit in late summer. *Highland Ringing Group Report* 3: 4-5.

SWANN, R.L. n.d.b. **611**
Primary moult of Redshanks in the Moray Firth basin. *Highland Ringing Group Report* 3: 16-20.

SWANN, R.L. and BAILLIE, S.R. 1979. **612**
The suspension of moult by trans-Saharan migrants in Crete. *Bird Study* 26: 55-58.

SWENNEN, C. 1977. **613**
Laboratory research on sea birds. Netherlands Institute for Sea Research, Texel.

TASKER, M.L. 1982. **614**
Moulting Shelducks on the Humber. *Bird Study* 29: 164-166.

TAYLOR, A.L. 1981. **615**
Adventitious molt in Red Knot probably caused by *Actornithophilus* (Mallophaga: Menoponidae). *J. Field Ornithol.* 52: 241.

THALER, E. 1979. **616**
Das Aktionssystem von Winterund Sommergoldhähnchen *(Regulus regulus* und *R. ignicapillus)* und deren ethologische Differenzierung. *Bonn. Zool. Monographien* 12.

THIEL, H. and VÖLKER, O. 1967. **617**
Zur Dauer der Federregeneration beim Waldkauz *(Strix aluco). J. Orn.* 108: 500-501.

THOMAS, D.G. and DARTNALL, A.J. 1971. **618**
Moult of the Curlew Sandpiper in relation to its annual cycle. *Emu* 71: 153-158.

THOMAS, D.K. 1977. **619**
Wing moult in Savi's Warbler. *Ringing and Migration* 1: 125-130.

THOMPSON, D.R. and KABAT, C. 1950. **620**
The wing molt of the Bob-white. *Wilson Bull.* 62: 20-31.

THÖNEN, W. 1965. **621**
Zur Schwingenmausser des Tüpfelsumpfhuhns. *Orn. Beob.* 62: 64-66.

THORNE, C.J.R. n.d.a. **622**
Moult of 12 species at Wicken Fen. *Wicken Fen Group Report* 6 (1974): 32-37.

THORNE, C.J.R. n.d.b. **623**
Primary moult of finches in Cambridgeshire. *Wicken Fen Group Report* 8 (1976): 26-29.

TIAINEN, J. 1981. **624**
Timing of the onset of postnuptial moult in the Willow Warbler *Phylloscopus trochilus* in relation to breeding in southern Finland. *Ornis Fenn.* 58: 56-63.

TICEHURST, C.B. 1938. **625**
A systematic review of the Genus Phylloscopus *(Willow-Warblers or Leaf-Warblers).* British Museum, London.

TINBERGEN, J.M. 1981. **626**
Foraging decisions in Starlings *(Sturnus vulgaris). Ardea* 69: 1-67.

TORDOFF, H.B. 1952. **627**
Notes on plumages, molts, and age variation of the Red Crossbill. *Condor* 54: 200-203.

TORDOFF, H.B. 1954. **628**
Further notes on plumages and molts of Red Crossbills. *Condor* 56: 108-109.

TORDOFF, H.B. 1962. **629**
Age and plumage of the Black Tern. *Kansas Orn. Soc. Bull.* 13: 7-8.

TORDOFF, H.B. and DAWSON, W.R. 1965. **630**
The influence of daylength on reproductive timing in the Red Crossbill. *Condor* 67: 416-421.

TREE, A.J. 1974a. **631**
Ageing and sexing the Little Stint. *Safring News* 3: 31-33.

TREE, A.J. 1974b. **632**
The use of primary moult in ageing the 6-15 month age class of some palaearctic waders *Safring News* 3: 21-24.

TREE, A.J. 1979. **633**
Biology of the Greenshank in Southern Africa. *Ostrich* 50: 240-251.

TRIAS, J., MARTÍNEZ, I. and LASURAIN, J. 1982. Notes sobre la muda del Boscarler Comu *(Locustella luscinioides),* al Delta de l'Ebre. *Butlletí del Grup Català d'Anellament 2* (1): 2-4. **634**

TUCK, L.M. 1972. **635**
The Snipes: a study of the Genus Capella. Canadian Wildlife Service Monograph Series No. 5.

TUCKER, J.J. 1975. **636**
White-winged Black Tern biometrics from Zambia. *Bulletin Zambian orn. Soc.* 7: 99.

TURCEK, F.J. 1966. **637**
On plumage quantity in birds. *Ekol. Polska* A, 14: 617-634.

TYSON, C.F. and PEPLER, G.R.M. 1976. **638**
Body moult of Reed and Sedge Warblers. *Radipole* 2 (1973-1975): 34-38.

ULLRICH, B. 1970. **639**
Ersatzbrut und Mauserbeginn beim Steinkauz *(Athene noctua). Vogelwelt* 91: 28-29.

VAN DIJK, A.J. 1980. **640**
(Observations on the moult of the Black-tailed Godwit *Limosa limosa*) *Limosa* 53: 49-57.

VAN HECKE, P. 1980. **641**
(Egg and wing measurements, body weight and moult of the Tree Pipit *(Anthus trivialis)*). *Vogelwelt* 101: 140-153.

VAN IMPE, J. 1978. **642**
(Moult migration and wing moult of the Pochard *(Aythya ferina)* in the Antwerp area). *Gerfaut* 68: 693-698.

VAN ORDEN, C. 1964. **643**
Laat ruiende Snor *(Locustella luscinioides). Limosa* 37: 316.

VAN ROSSEM, A. 1923. **644**
A study of some plumages of the Black Tern. *Condor* 25: 208-213.

VAN TYNE, J. and BERGER, A.J. 1976. **645**
Fundamentals of Ornithology. 2nd. Ed. Wiley, New York.

VAURIE, C. 1959. **646**
Birds of the Palaearctic Fauna: Passeriformes. Witherby, London.

VERBEEK, N.A.M. 1973. **647**
Pterylosis and timing of molt of the Water Pipit. *Condor* 75: 287-292.

VERBEEK, N.A.M. 1977. **648**
Timing of primary moult in adult Herring Gulls and Lesser Black-backed Gulls. *J. Orn.* 118: 87-92.

VERHEYEN, R. 1950. **649**
La mue du Coucou d'Europe *Cuculus c. canorus* L. *Gerfaut* 40: 212-231.

VERHEYEN, R. 1953. **650**
Exploration du Parc National de l'Upemba. Oiseaux. Institut des Parcs Nationaux du Congo Belge, Bruxelles.

VERHEYEN, R. 1956. **651**
Note sur la mue alaire et caudale chez les engoulevents, (Caprimulgidae). *Gerfaut* 46: 35-40.

VERHEYEN, R. 1956. **652**
La mue de la queue chez les hiboux et les chouettes (Striges). *Gerfaut* 46: 121-125.

VERHEYEN, R. 1957. **653**
Bijzonderheden over het jongkleed der Spechten (Picidae). *Gerfaut* 47: 177-182.

VERHEYEN, R. 1958. **654**
Note sur l'absence de la cinquième rémige secondaire (diastataxie) dans certaines groupes d'oiseaux récents et fossiles. *Gerfaut* 48: 157-166.

VERWEY, J. 1922. **655**
The moult of *Uria troille* (L.) and *Alca torda* L. *Ardea* 11: 99-116.

VERWEY, J. 1924. **656**
[Results of a study of the moult of the Guillemot, Razorbill and native species of divers]. *Tijdschr. ned. dierk. Ver.* (2) 19: vii-viii.

VERWEY, J. 1927. **657**
Maturity and breeding dress in birds. III and IV. The plumages of *Tringa canutus* L. and of *Tringa crassirotris* Temm. & Schlegel. *Zool. Meded., Leiden* 10: 2-28.

VERWEY, J. 1930. **658**
Geschlechtsreife und Prachtkleid der Vögel. V. Geschlechtsreife, Prachtkleid und Mauser, besonders bei einjähringen *Calidris canutus* (L.). *J. Orn.* 78: 234-245.

VEVERS, H.G. 1964. **659**
Colour. pp. 142-144 *in* THOMSON, A.L. (Ed.). *A new dictionary of birds.* Nelson, London.

VICKBERG, P. 1974. **660**
(Moulting records at Signilskär in the early autumns of 1971-72). *Lintumies* 9: 57-61.

VICKBERG, P. 1975. **661**
(Primary moult of irruptive Great Spotted Woodpeckers). *Lintumies* 10: 84-89.

VILLAGE, A., MARQUISS, M. and COOK, D.C. **662**
1980. Moult, ageing and sexing of Kestrels. *Ringing and Migration* 3: 53-59.

VLUG, J.J. 1974. **663**
Zomercontraties van de tuut *Podiceps cristatus* aan de triese kust tussen staveren en de lemmer. *Limosa* 47: 16-22*

VOITKEVICH, A.A. 1966. **664**
The feathers and plumage of birds. Sidgwick and Jackson, London.

VOLKER, O. 1957. **665**
Die experimentelle Rotfärbung des Gefieders beim Fichtenkreuzschnabel *(Loxia curvirostra). J. Orn.* 98: 210-214.

VOOUS, K.H. 1970. **666**
Moulting Great shearwater *(Puffinus gravis)* in Bay of Biscay. *Ardea* 58: 265-266.

VOOUS, K.H. 1977. **667**
List of recent Holarctic bird species. British Ornithologists' Union, London.

WALKER, A.F.G. 1970. **668**
The moult migration of Yorkshire Canada Geese. *Wildfowl* 21: 99-104.

WALLGREN, H. 1954. **669**
Energy metabolism of two species of the genus *Emberiza* as correlated with distribution and migration. *Acta Zool. Fenn.* 84: 1-110.

WALTERS, J. 1978. **670**
The primary moult in four gull species near Amsterdam. *Ardea* 66: 32-47.

WALTERS, J. 1979. **671**
The onset of the postnuptial moult in the Common Tern *Sterna hirundo* near Amsterdam. *Ardea* 67: 62-67.

WALTERS, J. 1980. **672**
Onset of moult in Avocets. *Bird Study* 27: 108.

WALTNER, M. 1976. **673**
Moult in Palaearctic waders. *Safring* 5: 14-16.

WARD, P. 1969. **674**
The annual cycle of the Yellow-vented Bulbul *Pycnonotus goiavier* in a humid equatorial environment. *J. Zool., Lond.* 157: 25-45.

WARD, P. and POH, G.E. 1968. **675**
Seasonal breeding in an equatorial population of the Tree Sparrow *Passer montanus. Ibis* 110: 359-363.

WASHINGTON, D. and HARRISON, C.J.O. 1969. **676**
Abnormal reddish plumage due to 'colour feeding' in wild Greenfinches. *Bird Study* 16: 111-114.

WATSON, A. 1957. **677**
The breeding behaviour, and food-ecology of the Snowy Owl *Nyctea scandiaca. Ibis* 99: 419-462.

WATSON, A. 1973. **678**
Moults of wild Scottish ptarmigan, *Lagopus mutus,* in relation to sex, climate and status. *J. Zool., Lond.* 171:207-223.

WATSON, G.E. 1962. **679**
Molt, age determination, and annual cycle in the Cuban Bobwhite. *Wilson Bull.* 74: 28-42.

WATSON, G.E. 1963. **680**
The mechanism of feather replacement during natural moult. *Auk* 80: 486-495.

WEAVER, R.L. 1942. **681**
Growth and development of English sparrows. *Wilson Bull.* 54: 183-191.

WEEDEN, R.B. 1966. **682**

Molt of primaries of adult Rock Ptarmigan in Central Alaska. *Auk* 83: 587-596.

WEITNAUER, E. 1977. **683**
Von der Mauser der Handschwingen beim Mauersegler *Apus apus*. *Orn. Beob.* 74: 89-94.

WELLER, M.W. 1976. **684**
Molts and plumages of waterfowl. pp. 34-38 *in* BELLROSE, F.C. *Ducks, Geese and Swans of North America*. Stackpole, Harrisburg.

WELTY, J.C. 1975. **685**
The Life of Birds. 2nd Ed. Saunders, Philadelphia.

WESTERSKOV, K. 1955. **686**
Notes on the post-juvenile moult and first-winter plumage in the Pheasant. *Br. Birds* 48: 308-311.

WESTERTERP, K., GORTMAKER, W. and **687**
WIJNGAARDEN, H. 1982. An energetic optimum in brood-raising in the Starling *Sturnus vulgaris*: an experimental study. *Ardea* 70: 153-162.

WESTPHAL, D. 1976. **688**
(Postjuvenile moult of the Greenfinch *Carduelis chloris*). *J. Orn.* 117: 70-74.

WETMORE, A. 1936. **689**
The number of contour feathers in Passeriform and related birds. *Auk* 53: 159-169.

WILKINSON, R. and AIDLEY, D.J. 1982. **690**
The status of Savi's Warbler in Nigeria. *Malimbus* 4: 48.

WILLEY, C.H. 1968. **691**
The ecology, distribution and abundance of the Mute Swan *(Cygnus olor)* in Rhode Island. Unpublished M.Sc. Thesis, University of Rhode Island.

WILLIAMSON, F.S.L. 1968. **692**
Calcarius lapponicus lapponicus (Linnaeus) Common Lapland Longspur. pp. 1597-1608 *in* AUSTIN, O.L. (Ed.). Life histories of North American Cardinals, Grosbeaks, Buntings, Towhees, Finches, Sparrows, and Allies. Order Passeriformes: Family Fringillidae. *U.S. Nat. Mus. Bulletin* 237.

WILLIAMSON, F.S.L. and EMISON, W.B. 1971. **693**
Variation in the timing of breeding and moult of the Lapland Longspur *(Calcarius lapponicus)* in Alaska, with relation to differences in latitude. *Bioscience* 21: 701-707.

WILLIAMSON, K. 1957a. **694**
Post-breeding moult of Crossbills. *Scott. Nat.* 69: 190-192.

WILLIAMSON, K. 1957b. **695**
The annual post-nuptial molt in the Wheatear *(Oenanthe oenanthe)*. *Bird-Banding* 28: 129-135.

WILLIAMSON, K. 1960a. **696**
Moult as a study in field taxonomy. *Bird Migration* 1: 171-175.

WILLIAMSON, K. 1960b. **697**
The development of young snipe studied by mist-netting. *Bird Study* 7: 63-76.

WILLIAMSON, K. 1961. **698**
Sequence of post-nuptial moult in the Starling. *Bird Migration* 2: 43-45.

WILLIAMSON, K. 1962. **699**
Post-nuptial moult in the Dunnock. *Bird Migration* 2: 114-115.

WILLIAMSON, K. 1965. **700**
Moult and its relation to taxonomy in Rock and Water Pipits. *Br. Birds* 58: 493-504.

WILLIAMSON, K. 1967. **701**
Identification for Ringers. 2. The Genus *Phylloscopus*. 2nd. Ed. *B.T.O. Field Guide* 8. British Trust for Ornithology, Tring.

WILLIAMSON, K. 1968a. **702**
Identification for Ringers. 1. The Genera *Cettia, Locustella, Acrocephalus* and *Hippolais*. 3rd. Ed. *B.T.O. Field Guide* 7. British Trust for Ornithology, Tring.

WILLIAMSON, K. 1968b. **703**
Identification for Ringers. 3. The Genus *Sylvia*. 2nd. Ed. *B.T.O. Field Guide* 9. British Trust for Ornithology, Tring.

WILLIAMSON, K. 1972. **704**
Reversal of normal moult sequence in the Spotted Flycatcher. *Br. Birds* 65: 50-51.

WILSON, E.A. 1910. **705**
The changes of plumage in the Red Grouse *(Lagopus scoticus* Lath.*)* in health and in disease. *Proc. zool. Soc. Lond.* 1910: 1000-1033.

WILSON, J.R., CZAJKOWSKI, M.A. and **706**
PIENKOWSKI, M.W. 1980. The migration through Europe and wintering in West Africa of Curlew Sandpipers. *Wildfowl* 31: 107-122.

WILSON, J.R. and MORRISON, R.I.G. 1981. **707**
Primary moult in Oystercatchers in Iceland. *Ornis Scand.* 12: 211-215.

WINKLER, R. 1975. **708**
Mauserverhältnisse bei Rauch- und Mehlschwalben auf dem Herbstzug. *Orn. Beob.* 72: 119-120.

WITHERBY, H.F. 1913. **709**
The sequence of plumage of the Rook. With special reference to the moult of the "face". *Br. Birds* 7: 126-139.

WITHERBY, H.F. JOURDAIN, F.C.B., **710**
TICEHURST, N.F. and TUCKER, B.W. 1938-1941. *The handbook of British birds*. Vols. I-V. Witherby, London.

WOOD, H.B. 1950. **711**
Growth bars in feathers. *Auk* 67: 486-491.

WOOD, J.B. 1976. **712**
The biology of Yellow Wagtails over-wintering in Nigeria. Unpubl. Ph.D. thesis, University of Aberdeen.

WOOD, N.A. 1974. **713**
The breeding behaviour and biology of the Moorhen. *Br. Birds* 67: 137-158.

WOOLFENDEN, G.E. 1967. **714**
Selection for a delayed simultaneous wing moult in Loons (Gaviidae). *Wilson Bull.* 79: 416-420.

WYNNE-EDWARDS, V.C. 1935. **715**
On the habits and distribution of birds on the North Atlantic. *Proc. Boston Nat. Hist. Soc.* 40: 233-346.

YOUNG, D.A. and BOAG, D.A. 1981. **716**
A description of molt in male mallards. *Can. J. Zool.* 59: 252-259.

ZEIDLER, K. 1966. **717**
Untersuchungen über Flügelbefiederung und Mauser des Haussperlings *(Passer domesticus* L.). *J. Orn.* 107: 113-153.

ZIPPELIUS, H.-M. 1972. **718**
Zur Brutbiologie des Pirols *(Oriolus oriolus)*. *Bonner zool. Beitr.* 23: 338-346.*

ZONFRILLO, B. 1982. **719**
Dark eye-patch of Fulmar. *Br. Birds* 75: 225-227.

GLOSSARY

Anterior: at or towards the front of the bird or a particular structure.

Arrested moult: a moult which would normally have been 'complete' which has stopped before all of the old feathers have been replaced. The old feathers, which have been retained, are then replaced in sequence at the time of the next complete moult. Thus a bird which has moulted P1 to P7 and has then arrested would not replace the outer primaries until after P1 to P7 had been renewed during the next moult. See also suspended moult (below).

Ascendant: moult which progresses from a point in the wing inwards, towards the bird's body.

Breeding plumage: plumage worn during all or part of the nesting season but which may be acquired much earlier. Some female ducks nest in a 'non-breeding' plumage but have earlier displayed and paired up in 'breeding' plumage.

Centre: a moult centre is the position in the wing or tail of a bird where its moult starts.

Centrifugal: a moult which progresses both ways, at the same time inwards and outwards, from a centre. For example tail moult which progresses from the central pair outwards.

Centripetal: the opposite of centrifugal — a moult which starts at two centres and converges. For example tail moult starting with the outer pair and finishing with the central pair of rectrices.

Complete moult: moult in which all of the feathers (including remiges and rectrices) are replaced. N.B. Some birds which have undergone a 'complete' moult may retain a few body feathers.

Cycle: a sequence of moult, generally of the primaries, which eventually leads to their complete replacement. In large birds a cycle may be interrupted and last more than a year, for such species active moult from different cycles may be present at the same time.

Descendant: moult which progresses from a point on the wing outwards, away from the bird's body.

Diastataxy: a term, mainly used by taxonomists, applied to the condition shown by species with a gap in the secondaries below the fifth greater covert: this gives the impression that the fifth secondary is missing. This is often very difficult to see in the field. The orgin of this condition is uncertain and is considered by some to be due to an 'extra' covert and by others to a 'missing' secondary (Humphrey and Clark 1961, Verheyen 1958).

Distal: towards the outside, far from the bird's body.

Eutaxy: where, unlike diastataxy, each secondary has a corresponding greater covert. This is the more common condition; all passerines are eutaxic.

Nuptial plumage: see breeding plumage.

Partial moult: a moult during which only some of a bird's feathers are renewed, the great majority of remiges and rectrices always being retained. It may affect almost all feathers of other tracts (e.g. post-juvenile moult of Robins *Erithacus rubecula* in the autumn) or very few (e.g. pre-breeding moult of some buntings *Emberiza* sp.).

Posterior: at or towards the back of a bird or particular structure.

Post-breeding moult: one which takes place after breeding (usually complete). In some species this moult overlaps part or all of the breeding cycle.

Post-juvenile moult: the first moult in which contour feathers are replaced, usually in the late summer of the first year and not affecting the flight feathers, but may be delayed until autumn of the second year in some non-passerines.

Pre-breeding moult: one which results in the assumption of 'breeding' plumage (generally a partial moult, not undertaken by all species).

Proximal: towards the inside, nearer the body — the reverse of distal.

Rectrix: main tail feather (plural rectrices).

Remex: main wing feather (plural remiges).

Score: moult scores are computed for a tract by adding together the scores for individual feathers within the tract (see fig. 7b, page 28). Thus the primary score (Pscore) of a bird with 10 primaries which are large enough to be recorded will lie between 0 (moult not started) and 50 (all feathers completely renewed).

Serially descendant: when a new moult cycle starts before the preceding one has finished (reached the outermost primary) so that there are two or more active moult centres in the wing which have started at different times and are following the same sequence. This is the equivalent of the German term *Staffelmauser* (Stresemann and Stresemann, 1966).

Simultaneous: all the feathers in a tract are lost at the same time.

Staffelmauser: 'step-wise moult', see serially descendant.

Suspended moult: a temporary interruption of moult which is subsequently resumed at the point of interruption (King 1972). Unless a bird is subsequently retrapped having resumed its moult after the interruption it is not possible to distinguish 'suspended' from 'arrested' (see above) moult. The term arrested is frequently used in the literature to refer to birds whose moult is more likely to have been suspended.

Tertial: the innermost secondary feathers, which usually moult as a group distinct from the other secondaries. In Part II the use of this term is generally restricted to Passerines as most have three easily found 'tertials'. They are also found in other groups but are seldom easy to define.

GERMAN TO ENGLISH GLOSSARY

Many papers on moult have been published in German and the short list below of German terms and their English equivalents may assist English-speaking readers using such German publications.

Absteigend .... Descending, decreasing
Adultkleid .... Adult plumage
  noch nicht in Adultkleid .... Immature (not yet in adult plumage)
Altvogel .... Old bird (Full adult)
Armflügel .... Secondaries (as a group)
Armschwinge .... Secondary
Aufsteigend .... Ascending, increasing
Brutkleid .... Breeding plumage
Diesjährig .... First year
  nicht diesjährig .... After first year
Dritten Jahr .... Third year
  nam dem dritten Jahr .... After third year
Feder .... Feather
Federgeneration .... Feather generation
Federwachstum .... Feather growth
Grossgefieder .... Quill (flight) feather
Handflügel .... Primaries (as a group)
Handschwinge .... Primary
Gefieder .... Feathers/Plumage
Jahreskleid .... Annual plumage
Jugenkleid .... Juvenile plumage
Jungvogel .... Young (juvenile) bird
Körper .... Body
Körpermauser .... Body (small feather) moult
Links .... Left
Mauser .... Moult
Mauserfolge .... Moult sequence
Mausernd .... Moulting
Mauserverlauf .... Moult progress
Mauserzeit .... Moulting season
Prachtkleid .... Breeding plumage
Rechts .... Right
Ruhekleid .... Non-breeding plumage
Schwanz .... Tail
Schwingen .... Wing
Schirmfedern .... Tertials
Sperlingsvogel .... Passerine
Schlichtkleid .... Non-breeding plumage
Sommerkleid .... Breeding plumage
Steuerfeder .... Tail feather, Rectrix
Teilmauser .... Partial moult
Vollmauser .... Full (complete) moult
Vorjährig .... Second year
  Alter als vorjährig .... After second year
Wachsend .... Growing
Wachstumsgeswindigkeit .... Speed of growth
Winterkleid .... Non-breeding (winter) plumage

RINGING EQUIPMENT FROM THE BTO

The National Bird Ringing scheme is run by the British Trust for Ornithology who are able to offer a very comprehensive range of ringing equipment. This is also available to other ringers operating abroad and much of the equipment is also very useful to other biologists. Some of the items are described below and full information and ordering procedures may be obtained from the Ringing Office, BTO, Beech Grove, Tring, Hertfordshire HP23 5NR. Numbered metal rings are only available to ringers who have proper qualifications under the BTO ringing scheme. The BTO does not currently supply colour rings.

MIST NETS

The use of mist nets is subject to legal restrictions in most countries. Within Britain they are only sold to suitably qualified BTO ringers and to the few biologists who have licences to catch bats. They are sold to ringers operating abroad who can provide the appropriate references from their own ringing schemes. Very good quality polyester netting is used. Fine material nets are available which have been completed in Japan but others, which have been finished in Scotland (North Ronaldsay, Orkney), are more durable.

SPRING BALANCES

Two makes of spring balance are available. Pesolas, from Switzerland, and Salters, made in Britain. The former, with metal parts, range in capacity from 5g to 20kg (larger ones available to special order) whilst Salters are made from plastic and have capacities from 50g to 1000g.

OTHER RINGING EQUIPMENT

Wing-rules: stainless steel rules (200mm and 300mm) with rivetted aluminium stops.

Ringing pliers: pliers purpose-made to suit British rings from 2mm to 19 mm internal diameter (including those made from very hard metals).

Bird-bags: cloth holding bags, with draw strings, for birds.

Head-lamps: torches which may be worn round the head so leaving both hands available for ringing operations.

Stationery: specially designed pads and notebooks for recording ringing information.

Cassettes: continuous cassettes for use with recorders for attracting birds. Specially designed amplifiers and all-weather loudspeakers are also available.

GUIDES

Four of the BTO series of guides (apart from this one) are of particular interest to the ringer. Three cover the identification of Old World warblers and are by the late Ken Williamson, the fourth, by Prater, Marchant and Vuorinen deals with Holarctic waders. The *Ringer's Manual* (issued free to licenced ringers in Britain) is available to ringers resident abroad and the indispensable *Identification Guide to European Passerines* by Lars Svensson is always available whilst it is in print.

SPECIES INDEX

Only bird species and groups are indexed. Page numbers in **bold type** refer to the species entries in Part II — The Species Accounts. An asterisk * denotes that an illustration relevant to the species is included. This index was compiled by Jill Graves.